# Geomembranes and the Control of Expansive Soils in Construction

# McGraw-Hill Construction Series: *M. D. Morris, Series Editor*

*Godfrey*
PARTNERING IN DESIGN AND CONSTRUCTION

*Matyas et al.*
CONSTRUCTION DISPUTE REVIEW BOARD MANUAL

*Peurifoy and Oberlander*
FORMWORK FOR CONCRETE STRUCTURES

*Rollings and Rollings*
GEOTECHNICAL MATERIALS IN CONSTRUCTION

*Palmer et al.*
CONSTRUCTION INSURANCE, BONDING, AND RISK MANAGEMENT

*Minsk*
SNOW AND ICE CONTROL MANUAL FOR TRANSPORTATION FACILITIES

*Nichols and Day*
MOVING THE EARTH—THE WORKBOOK FOR EXCAVATION, 4th ed.

# Geomembranes and the Control of Expansive Soils in Construction

Malcolm Steinberg

P.E., F.ASCE
Steinberg and Associates
El Paso, TX

McGraw-Hill

New York  San Francisco  Washington, D.C.  Auckland  Bogotá
Caracas  Lisbon  London  Madrid  Mexico City  Milan
Montreal  New Delhi  San Juan  Singapore
Sydney  Tokyo  Toronto

Library of Congress Cataloging-in-Publication Data

Steinberg. Malcolm L.
    Geomembranes and the control of expansive soils in conservation / Malcolm Steinberg
       p.   cm. — (Construction series)
    Includes bibliographical references and index.
    ISBN 0-07-061178-5
    1. Swelling Soils.   2. Geomembranes.   I. Title.   II. Series Construction series (McGraw Hill, Inc.)
    TA711. S74   1998
    624.1'5136—dc21                                                         98-24787
                                                                                                                                    CIP

# *McGraw-Hill*
*A Division of The McGraw·Hill Companies*

Copyright © 1998 by The McGraw-Hill Companies, Inc. All rights reserved. Printed in the United States of America. Except as permitted under the United States Copyright Act of 1976, no part of this publication may be reproduced or distributed in any form or by any means, or stored in a data base or retrieval system, without the prior written permission of the publisher.

1 2 3 4 5 6 7 8 9 0   FGR/FGR   9 0 3 2 1 0 9 8

ISBN 0-07-061178-5

*The sponsoring editor for this book was Larry S. Hager, the editing supervisor was Paul R. Sobel, and the production supervisor was Pamela A. Pelton. It was set in Palatino by Dina E. John of McGraw-Hill's Professional Book Group composition unit.*

*Printed and bound by Quebecor/Fairfield.*

McGraw-Hill books are available at special quantity discounts to use as premiums and sales promotions, or for use in corporate training programs. For more information, please write to the Director of Special Sales, McGraw-Hill, 11 West 19th Street, New York, NY 10011. Or contact your local bookstore.

---

Information contained in this work has been obtained by The McGraw-Hill Companies, Inc. ("McGraw-Hill") from sources believed to be reliable. However, neither McGraw-Hill nor its authors guarantees the accuracy or completeness of any information published herein and neither McGraw-Hill nor its authors shall be responsible for any errors, omissions, or damages arising out of use of this information. This work is published with the understanding that McGraw-Hill and its authors are supplying information but are not attempting to render engineering or other professional services. If such services are required, the assistance of an appropriate professional should be sought.

---

 This book is printed on recycled, acid-free paper containing a minimum of 50% recycled, de-inked fiber.

*To Elizabeth, my wife, friend, helper*

# Contents

Construction Series Preface  xiii
Foreword  xv
Acknowledgments  xvii

**Chapter 1. Introduction**  1

    1.1  The Worldwide Challenge  1
    1.2  Expansive Soils Studies  3
    1.3  Solving Expansive Soils Problems  4
    1.4  Geomembrane Technology  5
    1.5  References  6

**Chapter 2. Geomembranes**  9

    2.1  Terminology  9
    2.2  Historical Development  11
    2.3  Applications of Geomembrane Technology  12
    2.4  Fabrics and Fabric Manufacturers  13
    2.5  Geographical Distribution of Geomembrane Use  14
    2.6  The Capabilities and Limitations of Geomembranes  14
    2.7  The Effectiveness of Geomembranes  15
    2.8  References  16

**Chapter 3. Expansive Soils**  17

    3.1  Expansive Soil Terminology and Physical Properties  17

3.2 Geographical Distribution of Expansive Soils   18
3.3 Damages Caused by Expansive Soils   21
3.4 Environmental Impact of Expansive Soils   21
3.5 Financial Impact of Expansive Soil Damage   22
3.6 Costs of Expansive Soil Disasters Compared with Other Natural Disasters   23
3.7 Long-Term Benefits of Studying Expansive Soils   23
3.8 References   26

## Chapter 4. Responses, Studies, and Theories    27

4.1 The Atterberg Tests   27
4.2 Continuing Studies by Universities   28
4.3 Texas Department of Transportation Studies   31
4.4 Federal Highway Administration Studies   32
4.5 Transportation Research Board Meetings and Publications   34
4.6 Suction Testing   34
4.7 Texas A&M University Studies   36
4.8 Ongoing International Studies   37
4.9 References   39

## Chapter 5. Facilities Affected    41

5.1 Types of Facilities   41
5.2 Highways, Roads, and Streets   43
5.3 Airfields   44
5.4 Railroads, Canals, and Pipelines   44
5.5 References   45

## Chapter 6. Testing    47

6.1 The Importance of Adequate Testing Prior to Design   47
6.2 Resources for Information on Soil Testing   48
6.3 On-Site Soil Inspection   50
6.4 Determining the Number of Test Sites and the Methods and Evaluating the Results   50
6.5 Approaching the Design and Construction Stages from the Test Phase   52
6.6 Selecting a Geomembrane   53
6.7 References   54

## Chapter 7. Design    55

7.1 General Design Considerations in Controlling Expansive Soils   55
7.2 Expansive Soils Control Options   58
7.3 Geomembrane Placement Design Considerations   62

Contents       ix

    7.4  Applicability and Feasibility of Various Design Approaches   64
    7.5  References   66

## Chapter 8. Case Studies: Buildings   69

    8.1  Expansive Soil Experiences in the United States, South Africa, Canada, Australia, and Israel   69
    8.2  New Construction and Remediation   75
    8.3  New Developments Using Geomembranes   83
    8.4  References   84

## Chapter 9. Case Studies: Buildings without Geomembranes   85

    9.1  Class-Action Suits in California, Colorado, and Texas   85
    9.2  Managing Expansive Soils around Large Institutional and Commercial Buildings   87
    9.3  More Than 2000 U.S. Buildings Damaged by Expansive Soils   89
    9.4  Preventive Strategies   89
    9.5  References   91

## Chapter 10. Case Studies: Texas Highways, Roads, and Streets   93

    10.1  Lessons Learned from Projects in Colorado and South Dakota   93
    10.2  Texas Highway Swelling Soil Projects   95
    10.3  An Assessment of the Texas Geomembrane Projects   117
    10.4  References   118

## Chapter 11. Case Studies: Highways, Road, and Streets beyond Texas   121

    11.1  Wyoming   121
    11.2  Other States   126
    11.3  Australia   130
    11.4  Israel   137
    11.5  An Evaluation   139
    11.6  References   139

## Chapter 12. Case Studies: Airports, Canals, Railroads, and Pipelines   141

    12.1  Airports   141
    12.2  Canals   142
    12.3  Railroads   144

12.4 Pipelines  145
12.5 References  146

## Chapter 13. Placement                                          149

13.1 Equipment Used in Remedial Treatments without Geomembranes  149
13.2 Horizontal Geomembrane Placement Methods  151
13.3 Vertical Geomembrane Placement Methods  151
13.4 Geomembrane Trench Backfill Methods  152
13.5 Australian Backfill Methods  153
13.6 References  154

## Chapter 14. Costs                                              155

14.1 Factors Affecting Costs  155
14.2 Cost Variations on Texas Highway Projects  156
14.3 Accuracy of Project Specifications  157
14.4 Comparison of Australian and U.S. Costs  158
14.5 Costs of Various Geomembrane Placement Rates  160
14.6 Costs of Various Treatment Methods  161
14.7 References  163

## Chapter 15. Durability                                         165

15.1 Justifiable Concerns  165
15.2 Tests by Manufacturers  166
15.3 Tests by State Agencies  167
15.4 Australian Experiences  168
15.5 Positive Indications of Geomembrane Durability  169
15.6 References  169

## Chapter 16. Last, But Definitely Not Least                     171

16.1 A Partial List of Dos  171
16.2 A Partial List of Definitely Don'ts  173
16.3 Further Dos and Don'ts  174
16.4 Conclusion  175

## Appendix A. Vertical Moisture Barrier                          177

1. Description  177
2. Materials  177
3. Physical Requirements  178
4. Packaging Requirements  179
5. Labeling or Tagging  180
6. Sampling and Testing Requirements  180

Contents

       7. Basis for Rejection   181
       8. Measurement and Payment   181

## Appendix B. Impermeable Plastic Membrane   183

       Section 630: Impermeable Plastic Membrane   183
       Section 631: Slotted Drains   185

## Appendix C. Special Provision for Impermeable Plastic Membrane   189

       Description   189
       Materials   189
       Construction   190
       Method of Measurement   191
       Basis of Payment   191

## Appendix D. Geomembrane   195

       Description   195
       Material Requirements   195
       Construction Requirements   196
       Weather Limitations   196
       Equipment   197
       Subgrade Preparation   197
       Geomembrane Placement   197
       Placement and Compaction of Aggregate   200
       Method of Measurement   200
       Basis of Payment   201

## Appendix E. Geotextile Fabric   203

       907-714.13 Geotextile Fabrics   203
       907-486 Pavement Fabric   205
       907-714.14 Geotextile Fabric for Moisture Barrier   206

Index   215

# Construction Series Preface

Construction is America's largest manufacturing industry. Ahead of automotive and chemicals, construction represents 14 percent of this country's Gross National Product. Yet it is unique in that it is the only manufacturing industry where the factory goes out to the point of sale. Every end product has a life of its own, and is different from all others, although component parts may be mass produced, or modular.

Because of this uniqueness, the construction industry needs and deserves a literature of its own, beyond reworked civil engineering texts, and trade publication articles.

Whether management methods, business briefings, or field technology, it will be covered professionally and progressively in these volumes. The working contractor aspires to deliver to the owner, a superior product ahead of schedule, under budget, and out of court. This Series, written by constructors, and for constructors, is dedicated to that goal.

*M. D. Morris, PE*
Series Editor

# Foreword

For one who deals with the problems of expansive soils in the last 30 years, I found the book most worth while reading for all geotechnical engineers as well as architects and contractors. We all realize that water is the main, if not the only cause of foundation failure. Moisture barrier is the most positive method in preventing water from entering the foundation soils. Nevertheless, the design and the installation of a moisture barrier are open to discussion

Engineers have different opinions as to the location, the depth, the extent as well as the type of material used for the barrier. Above all, the cost of the installing moisture barrier should be taken into consideration. This book provides the answers.

I have read papers and publications on the theory of expansion, the mechanics of swelling pressure, and the various methods of treatment, yet I have to find a positive solution. Reading this book is a breath of fresh air. At last I know where and how to find a practical method in dealing with moisture barrier. I could be at Rock Springs, Wyoming, inspecting a canal at Bombay, India or designing the huge "South to North" water transfer project in China, the answer is to be found in this book.

This book deals not only with method of installation, but also record the behavior of the barrier many years after completion. To a geotechnical engineer, long term monitoring is probably the most important cost control element.

*Fu Hua Chen*
honorary member ASCE

# Acknowledgments

My thanks to a whole host of folks: My coworkers at the Texas Highway Department, now the Texas Department of Transportation, including: Gerald Hewitt, Raymond Stotzer, John Nixon, Kathleen Jones, Charles Baxter, Pablo De Arkos, Dale Stein, Michael Taylor, Hal Stanford, Gene Clement, Robert Magers, Robert Kovar, Consuela Flores, Blythe Lowery, Henry Hardy, Thomas Walthall, and a whole host more.

Those from Manufacturers Universities the Transportation Research Board, federal, state, city agencies, and consulting engineers around the country: William Hawkins, Fu Hua Chen, Nicholas Quarm, Steve Raabe, Robert L. Lytton, Thomas Scullion, George Osborne, Roy Olson, Charles Shambelan, Roger Simienko. Guy Hopp, Gordon McKeen, Johm Nelson, Harry Tan, Ronald Hudson, Alice Comer, Gordon Watt, Michael Hager, Joni Kachelhoffer, G. Riedl and Jim Miller, Art Greengard, Robert Thompson, Alvin Kaplan, Donald Snethen, Lawrence Johnson, Richard Lueck, Robt. Browning, Doug Smith, Paul Teng, Jerry Bowman, Douglas Steadman. Juan Sandoval, Chester Drash, William Henry, Dave Cochran, and Arvid Noori, J. Cotecson, Al Spensko and many more.

From our engineers across the seas: Gdalyah Wiseman, Amos Komornick, Moshe Livneh, Shimon Neshichi, Jocob Uzan, Sam Frydman, Uri Ackerbaum, James Holden, Kerry McManus, Robert Evans, Sandra DeMarco, and Anthony Mann.

And certainly not least our editors Larry Hager and Doc Morris.

# 1
# Introduction

The threat to public safety and the potential for destruction of property caused by unconstrained expansive soils are growing exponentially. In the United States alone, the cost of damage from expansive soils is $10 billion annually. In China, the cost of damage to the railroad system has been estimated at 100 million yuan (Chinese currency) 12 million U.S. dollars a year. Many techniques such as the use of geomembranes have been tried in an effort to control the expansive soils in construction projects.

This book describes and defines expansive soils and their physical properties. The book also explains how and why expansive soils swell and shrink, as well as the conditions in which they become destructive. Successful and unsuccessful efforts to control the destructive movements of these soils in actual construction cases are described. Situations in which geomembranes have been used are discussed, with an evaluation of their success, costs, and legal concerns.

The practical information that this book provides on geomembranes is not overly theoretical and should prove helpful to engineers, builders, designers, and property owners.

## 1.1 The Worldwide Challenge

Expansive soils have been found on all the continents with the exception of the polar continents. Seven international conferences have brought forth papers of pertinent case studies from around the world. They include reports on damage to a Saudi Arabian king's palace to private homes in the United States from the Texas Gulf coast to the Imperial Valley of

California. Damages to subdivisions in the United States have resulted in class-action suits in Colorado and California. Individual home owners' suits have appeared in courtrooms in Corpus Christi, Texas, on the Gulf coast, to San Antonio, 160 miles inland. Similar destructive results from these soils have been reported in Burma, Australia, the United States, South Africa, India, Canada, China, and Israel.

Transportation facilities have been similarly affected by these destructive soil movements. More than one dozen state transportation agencies in the United States have had to aggressively address the damage to their highways by using geomembranes on more than 100 projects. The use of geomembranes has usually been evaluated as successful. One probing study conducted by the U.S. Army Corps of Engineers (COE), the Waterways Experiment Station (WES), indicated that probably 40 of the 50 states have these expansive soils. Around the world, reports confirm that the expansive soil challenge is not faced by the United States alone. China indicates that almost 20 percent of its railway system is built across swelling soil subgrades. Fu Hua Chen notes nine Chinese provinces affected by swelling soils, located usually between Longitudes 22 and 28 [1].

India, Chen records, has large areas of the expansive clay, a black cotton soil called *Regar*. The clay covers almost the entire Duccan Plateau, an area estimated to be 200,000 square miles. Chen quotes Krishnan as further identifying the black cotton as clayey to loamy with a variety of chemicals. Singh is noted by Chen as identifying the montmorillonitic presence where the total black cotton area is on India's western shore. In the Middle East, Chen records highly expansive soils in northwestern Jordan causing cracking in building walls. In Saudi Arabia, again it's the northwestern sections of the kingdom where Chen identifies the presence of expansive clays. The Medina area has montmorillonites while mudstone and claystone are prevalent in other parts of the area. Chen notes that the claystone is similar to that observed in Colorado's front range. Israel has considerable areas of expansive soils in its small country. Chen credits Professor J. G. Zeitlin's early work there in identifying these soils. They appear to swell, heave, and shrink, affecting small and large buildings as well as transportation facilities.

In Africa, Chen observes the presence of the expansive soils in Sudan, which extend through more than one-third of the country's area. Some of the deposits, including montmorillonites, come from deposits of the Ethiopian highlands. Ethiopia has significant amounts of these expansive soils largely identified as black clays in the eastern part of the land. Zimbabwe (formerly Rhodesia) has the expansive soils in their western and southern reaches; Chen credits the work of C. F. Van der Muerwe and M. Aronowitz, for this discovery, which they made using cation exchange

Introduction

capacity (E/C) for estimating the reactivity of swelling clays. In South Africa, Chen indicates that the challenge of the expansive soils was noted in the early 1950s. Their first symposium on expansive clays was held in 1957, and it was sponsored by the South African Institute of Civil Engineers.

In Europe experiences with the expansive soils are noted by Chen in England and in Spain. The French have faced a similar challenge in their railroad work. In the Americas voluminous studies have been conducted in the United States, for example, the Waterways Experiment Station's report. Canada and Mexico are conducting substantial studies as well. South American engineers have addressed and written on the expansive soils problems in Peru, Brazil, Colombia, and Venezuela. Australian studies have shown that their most heavily populated areas are the ones with the most expansive soils. They are currently making substantial progress in controlling this challenge.

An examination of the reported occurrences of the expansive soils indicates the magnitude of the challenge worldwide. What is perceived is, in a sense, just the beginning. The problem has been exacerbated by the development of human engineering projects and the increasing number of people populating the earth. Minimal human facilities did not feel the impact of the expansive soils. When Abraham led his tribe from Ur to Israel, their tents and camel paths were not impeded by the expansive soils they crossed. When Native Americans chased the buffalo or moved their tepees or hogans, the expansive soils troubled them not at all. Today demands are made for far more sophisticated structures and transportation facilities, all of which are subject to the destructive movements of these expansive soils.

When buildings in Mandalay, Burma, suffer damages or a highway between San Antonio and Seguin needs extensive repair, and when this type of damage is repeated in Brazil, Canada, Australia, Africa, and the United States, it is apparent that the challenge is growing with each passing moment. Before engineers, contractors, developers, and planners get down to the construction stages of new facilities, adequate soil investigations must be made. If they are not approached with proper professional consideration, the movements of the expansive soils will result in the destruction of the facilities.

## 1.2 Expansive Soils Studies

Efforts over the decades to understand these soils began in the early days of the twentieth century in the emerging discipline of civil engineering known as *soil mechanics*. Dr. Karl Terzaghi, professor of engineering practice at Harvard University, published a treatise in 1925, which is acclaimed as the first definition of the civil engineering art of soil mechanics. Early studies focused on the destructive expansive actions of certain clays and shales. Henry Porter, research engineer for the Texas Highway Department (THD), conducted studies in Navarro and Guadalupe counties, Texas, in the expansive soils there. This work for the THD was reported by Porter at the International Conference on Soil Mechanics and Foundation Engineering (ICSMFE) at Harvard University in 1936. His later publications included a series of bulletins by Texas A&M University, formerly the Texas Agricultural and Mechanical College, in 1942. The focus was on the impact of swelling clays on highways and the efforts to control the impact. D. Wooltorton, at the same 1936 Harvard meeting, reported on destructive movements to almost 100 structures in the Mandalay area of Burma. His investigation found the damages resulted not from structural design shortcomings or faulty construction but rather from expansive soils. During the discussion following Wooltorton's presentation, Willard Simpson, Sr., a San Antonio, Texas, consulting engineer, reported that he had been studying these swelling clays for the preceding 20 years. Simpson, a 1905 graduate of the Massachusetts Institute of Technology (MIT), worked and consulted with Terzaghi in dealing with the problems in foundation design caused by the expansive soils. During the 1940s increased attention to the swelling clays related to the construction boom of military bases, particularly in Texas. Conard Kelly, who had been a Texas Highway Department district laboratory engineer, began using lime to reduce the expansive characteristics of the swelling clays. In Europe Leo Casagrande used electroosmotic stabilization of a clay for German submarine pens. There was some effort by the University of Arizona and the Arizona Department of Transportation (AZDOT) to replicate these efforts. The results indicated that the high cost of the control methods made them uneconomical. Following World War II the use of lime to stabilize the clay, or attempt to do so, increased. This was the beginning of a more intense period of studies to find methods to control these expansive soils.

## 1.3 Solving Expansive Soils Problems

Some of the earliest efforts focused on *reducing the plasticity index* (PI) *of the clay.* It became apparent with climatic conditions and chemical characteristics of the clays and shales with high plasticity indices that the result was considerable soil movements. The percent of lime added to the soil was increased with the higher PIs. Initially it was recommended to lime stabilize the top 6 inches of the subgrade. Later the depth increased to 1 foot and 2 feet. In Belgium they recorded a treatment of 5 feet. *Electroosmosis* was tested in Arizona. Replacement of the expansive soil with nonexpansive material is a solution used particularly on transportation facilities. *Ponding* has been used on several projects following the early work by Porter. Avoiding the expansive soil remains an attractive option. Regrettably, this option becomes harder to implement as land development becomes more intensive worldwide.

Another solution developed was *encapsulization*. This frequently involved the use of heavy sprayed asphalt applications. The Houston Urban District encapsulated an embankment to maintain a uniform moisture content in the 1950s. The Colorado Highway Department, under research engineer B. A. Brakey, used a heavy application of sprayed asphalt of a gallon per square yard to minimize moisture change in an expansive Mancos shale. About that time the Du Pont Company began producing a geomembrane, Typar, which was offered to replace the heavy asphalt application. As a postscript, Paul Teng, Mississippi Highway Department research engineer now with the U.S. Department of Transportation, used the heavy asphalt application. A year later, going back to the facility, he excavated to find the asphalt, but he could not locate any.

## 1.4 Geomembrane Technology

The Du Pont Company's Typar was the first geomembrane used in the United States by a state transportation department. It was placed horizontally in the 1960s on one of Delaware's farm roads. It was the first of over 100 transportation projects in the United States to use the geomembranes. Other manufacturers soon began providing the product. Horizontal placement was followed by vertical placement, and then both horizontal and vertical. It is being used successfully in the United States, Australia, and Israel, and trial use has so far been successful in Europe.

In the United States, over 100 transportation-related highway, road, and street projects have used geomembranes over expansive soil subgrades. In addition to the early geomembrane highway test by the Colorado DOT, a vertical placement was used on a South Dakota highway rehabilitation. The Texas Department of Transportation (TexDOT) (formerly the Texas Highway Department) began its geomembrane projects with the horizontal placement of the geotextile on an urban arterial street. *Deep vertical fabric moisture barriers* (DVFMB) followed on several interstate highway rehabilitation projects. TexDOT now has almost 25 such projects. Not every one has been a success. Pavement roughness has recurred, and pavement cracking has appeared on some of the projects. However, it has succeeded in reducing the destructive movements of the expansive soils. It has repeatedly led to a reduction in maintenance costs, in drivers' and legislators' complaints, and improvement of the cash crunch.

Wyoming's Department of Transportation (WyDOT) followed in the early 1980s with their first of over 50 geomembrane projects. Originally, they placed the geomembrane horizontally, but later they placed the geotextile horizontally and vertically. Their evaluation of the geomembrane is similar to that of Texas. Usually it does a great job. The degree of their satisfaction can be gauged in the fact that they are using over 4 million square yards of the geomembrane. A dozen other states have used the geomembrane similarly. The geomembrane has been used as a moisture barrier in expansive soil situations with rail lines in China and France, airfields, and building developments in Australia and the United States. Geomembranes are not an innovation anymore. There's a track record you can draw upon. There are considerable expansive soil areas just waiting for those geomembranes to control the swelling and the shrinking.

The Du Pont Company in the United States was the first to offer Colorado, then the Texas Highway Department, a geomembrane to use on their highways in the effort to control these expansive soils. Conversations with one of their highest-ranking research scientists in the United States, Dr. Roger Siemienko, indicates early use of the geomembranes in Europe may have been to strengthen dykes in Holland. Its major use currently may be for carpet backing, but its use in China on roads and railroads is really taking off. Diapers have been claimed to be the biggest users of the geomembrane product.

Dr. Guy Hoop, chief of technical management, Du Pont—Luxembourg, indicates further breadth on the use of geomembranes around the world. He mentions that they are used in 140 km (94 miles) on Russia's railroad line, the French high-speed trackage, and Saudi airfields.

## 1.5 References

1. Fu Hua Chen. *Foundations on Expansive Soils.* Elsevier, Amsterdam, 1988.

2. H. C. Porter. "Roadway of Runway Soil Mechanics Data." Texas Highway Department, Austin, 1942.
3. D. Wooltorton. "A Preliminary Investigation into the Subject of Foundation in the `Black Cotton' and `Kyatti' Soils of the Mandalay District, Burma." *Proceedings of the International Conference on Soil Mechanics and Foundation Engineering*, Harvard University, Cambridge, MA, 1936.
4. W. E. Simpson. "Settlement Records of the Gulf Building." *Proceedings of the Seventh Texas Conference on Soil Mechanics and Foundation Engineering*, Austin, 1947.
5. C. M. Kelly. "A Long Range Durability Study of Lime Stabilized Base." National Lime Association, Arlington, VA, 1977.

# 2
# Geomembranes

This chapter provides some working definitions with which to consider geomembrane technology. Also discussed are the historical developments that led to the invention and use of geomembranes, their chemical makeup and properties, their applications in construction projects, the fabrics they are made of as well as the manufacturers of those fabrics, the geographical locations of their use, and their capabilities and limitations.

Geomembranes have been in use around the world for several decades, with impressive results. Granted, they won't solve all the world's problems, but they will go a long way toward preventing expansive soil disasters.

## 2.1 Terminology

As defined by the American Society of Testing Materials (ASTM D 4439), a *geomembrane* is a very low permeability synthetic liner or barrier that has been coated with a geotechnically engineered material to control fluid migration in a human-made structure or system. Professor Robert Koerner's book *Designing With Geosynthetics* states that geomembranes are usually made from very flexible continuous polymeric sheets. Geomembranes can also be made by impregnating geotextiles with elastomer sprays or bitumen composites.

Koerner identifies 7 basic types of geomembranes:

Chlorinated polyethylene

Chlorosulfonated polyethylene

Ethylene interpolymer alloy

High-density polyethylene

Polypropylene

Polyvinyl chloride

Low-density polyethylene

Of the 14 possibilities, 11 are available with or without scrim reinforcement. *Scrim* is a lightweight open-weave fabric with a low mass per unit area, which is used to reinforce a geomembrane. Sometimes the scrim is made thinner by calendering it before it is blended with a compound. A scrim is characterized by its count and the linear density of its yarns. The *count* is the number of yarns per unit of width. The *linear density* of a yarn is its mass per unit length. The *mass per unit area* of a scrim is derived by multiplying the count by the linear density in both directions [1].

The basic elements making up these fabrics are carbon, hydrogen, and oxygen. The nonwoven fabrics are porous textile materials. They are usually manufactured as flat sheets composed primarily or entirely of fibers assembled in webs. The fibers are categorized as *bonded, formed,* or *engineered*. They are manufactured by means other than spinning, weaving, or knitting. *Spun-bonded fibers* and *webs* are made simultaneously directly from bulk polymers. The common base polymers are polypropylene, polyethylene, polyester, and nylon. Bonding is accomplished through mechanical, friction, or chemical reaction or with adhesives, either latex, sprays, or thermal.

The geotextile is porous to water but not to soil fines. It is because of this characteristic that it has been used as a separator on road work. It has also been used as a nonwoven fabric separator on roadway drainage projects [2]. Both the high- and low-density polyethylene geomembranes are available in rolls. Widths from 3 m (10 feet) to 10.5 m (33 feet) with lengths to 300 m (1000 feet) are typically available. Sheet thicknesses may vary from 1.0 to 2.5 mm (30 to 75 mils), although thinner geomembranes have been satisfactorily supplied and used by a variety of manufacturers, and some of the Typar geomembranes have ranged from 7 to 9 mils thick. Of the 12 products available from six manufacturers, only 1 is 60 mils thick, with the rest being 24 mils or less. The thicker geomembranes are usually supplied in large panels prefabricated in factories, and their seaming usually ranges from 1.4 to 2.5 m (4 to 8 feet) wide.

The geomembranes are relatively impermeable materials. In fact, their permeability is substantially lower than that of compacted clay, which usually places them in the *impermeable* category. J. P. Giroud in 1984 noted that, when layered, they can be used to construct an impermeable structure. When a geomembrane is used as a barrier for hazardous or toxic wastes in containment structures, as it frequently is, leakage is the more critical characteristic. However, since this book deals primarily with structures and expansive soils, leakage will not be considered in as much detail as it would be otherwise.

## 2.2 Historical Development

A *polymer* is a molecule with many units; in contrast, a *monomer* is a molecule with only a single unit. Polymers are made from monomers in a reaction called *polymerization*. An example is a polyvinyl chloride polymer, which is made from a monomer, vinyl chloride [3]. Polyvinyl chloride (PVC) was first prepared in Germany in 1835, but it was not produced commercially until 1935. Progress in the production of PVC waited upon the development of a plasticizer. Typical PVC formulations are made up of 58 percent PVC, 33 percent plasticizer, and 19 percent stabilizer. One of the first plasticizers was ethyl vinyl acetate, which was used in some of the early geomembrane projects.

Polymeric geomembranes are not absolutely impermeable, but, compared to geotextiles and clay, they are relatively impermeable. Permeability is measured by water vapor transmission tests. The results of these tests are used to classify geomembranes as barriers to liquids and vapors. Primary polymers used in geosynthetic materials are olefins, polyesters, polystyrene, and nylons. *Olefins* are used as the base for many of the geomembranes. They are made from long chains of synthetic polymers, and they are at least 85 percent (by weight) ethylene, propylene, or other olefin unit. In addition to being chemically inert and lightweight, they have high strength and abrasion resistance. Increasing the linearity of the polymer structure increases the strength and density of the polymer.

The geomembranes are polymers of either woven or nonwoven materials. The woven fabrics are made by weaving filaments, yarns, or ribbons of slit plastic film. The woven fabrics, have varying structural features and thus perform differently, and some are more tightly woven than others to permit only a small percentage of open surface area. To facilitate their use as a drainage material, the areas of open surface are increased. To minimize the movement of the woven fibers or ribbons, they must be immobilized. Their use as a barrier for moisture movement requires the minimization of the areas of open surface.

The first nonwoven fabric textile patent was issued in 1936. The nonwoven fabrics can be made in a variety of ways. Short chopped fibers or continuous filaments can be used. They are composed of needle-punched fabrics, which are made by repeatedly passing barbed needles through bats of staple or continuous filaments. This causes the nonwoven fabrics to become further entangled, increasing their dimensional stability. Other nonwoven fabrics are bonded with an adhesive resin. The resin becomes a source of strength that coats and holds the fibers together. Some nonwovens are classified as spun bonded while others are heat or melt bonded. In the latter cases, bonding develops from heat and pressure [4].

Fibers have a long history as a construction material. Mongols used sheeps' wool as a saddle blanket; it became dense over time. They also used it as a cover for their dwellings, the yurt. As early as 2500 B.C., nonsoil uses included wood as a corduroy road. In 1935, cotton fabric was used on a South Carolina road. Florida used automobile upholstery fabric as an erosion control measure in 1950. This led to geotextiles being used in roads around the world for erosion control. Modern era geotextile work began in Europe in the 1960s. Rhone Poulence manufactured a needle-punched fabric called BIDIM, and ICI developed TERAM (Celanese/Mirafi), a nonwoven spun-bonded sheath core. Du Pont's Typar, now a REEMAY product, was a spun-bonded polypropylene followed by Phillips Petromat, currently an Amoco–owned geomembrane, a needle-punched polypropylene.

## 2.3 Applications of Geomembrane Technology

Geomembranes were developed and used in Europe in a multitude of situations. The Dutch used them extensively in dyke surge protection construction in North Sea areas beginning in the 1950s. In the 1960s, the Du Pont Company in the United States manufactured a geomembrane product. It was made of polypropylene, which was spun bonded and coated with an ethyl vinyl acetate (EVA) to give it an impermeable surface. Du Pont called the geomembrane *Typar.* Dr. Harry Tan, the Du Pont Company's geotechnical consultant, viewed Typar as a means to control the moisture changes in an expansive soil.

Most synthetic materials used for making geotextiles, such as geomembranes, have limited resistance to prolonged exposure to sunlight. They do, however, resist rot, mildew, other microorganisms, and chemicals normally found in the ground and water. Polypropylene resists prolonged exposure to acids or alkyls better than nylon or polyester.

A common use of polyvinyl chloride geomembranes is in the construction of containment vessels, waste ponds, and landfills. Also, in the United States, Australia, and Israel, geomembranes have been used in highway projects to control the destructive movement of expansive soils. They have also been used on city streets and airfields in the United States. The U.S. Bureau of Reclamation included geomembranes in their canal designs. Pipelines in the United States and Israel have used geomembranes in their efforts to prevent expansive soil damage.

China and France are using geomembranes successfully in rehabilitating elements of their railroad lines. Significant projects have incorporated geomembranes to control expansive soils affecting both residential and commercial units in the United States and Australia. The use of geomembranes is increasing because they can provide an economical solution to the hazards of expansive soils.

The availability of several yarn types and combinations of them has resulted in the proliferation of manufacturers who also serve as contractors, subcontractors, and fabricators in supplying geomembranes. Some use virgin materials while others use recycled materials. Using recycled materials has become a more common approach in Europe than in the United States. In Germany, using recycled materials is required by law. In the United States companies use recycled plastic bottles and other consumer-used plastic, hydrocarbon products, and even x-ray film to produce geomembranes.

## 2.4 Fabrics and Fabric Manufacturers

The first fabric geomembrane to be manufactured and introduced in the United States was called *Typar*. Its horizontal placement on the Colorado highway system was the first use of a geomembrane on a transportation facility in the United States. When the first two contracts with the Texas Department of Transportation (TexDOT), formerly the Texas Highway Department, were written, Du Pont's Typar was the only geomembrane available. However, it wasn't long before there was considerable competition for the geomembrane projects. The Phillips Petromat MB (moisture barrier), now part of Amoco Fabrics, and Mirafi MCF 1212, now Nicolon/Mirafi, followed in rapid succession. Other manufacturers later included Polyflex, Fabrene, Lorotex, Midwest Canvas, Dane Chemco, Novastrene, and West Industries. In some cases one firm manufactures the geotextile, while another supplies and applies the waterproofing agent. The controlling element is the requirement for the finished product, which must meet the specification the contracting agency has required.

In many cases the style and/or type numbers have varied as have the properties of the different manufacturers. Typar styles, now manufactured by REEMAY, originally ranged from 3153, 3313, 3353, 3401, and T-063. The

Phillips MB geomembrane, now manufactured by Amoco, is another example of the changing scene. Weights of the geomembranes used on state highway projects in efforts to control the expansive soils have varied from 2.4 to 6.5 oz/yd$^2$. Thicknesses have varied from 7 to 60 mils. In many of the first Texas highway projects the geomembrane was 10 feet wide and rolls were 300 feet long. Manufacturers now talk of geomembrane rolls being 22 feet wide.

## 2.5 Geographical Distribution of Geomembrane Use

Geomembranes have been used in efforts to control expansive soils on many of the world's continents, especially the North American continent. In Europe, the Dutch used a form of geomembrane material to reinforce their system of dykes, and the Germans have used them to line their waste ponds. In Asia, the Israelis have used geomembranes on their road system, and the Chinese have used geomembranes placed horizontally to correct problems on their rail system. The Australians have similarly used geomembranes on both their highway and street systems as well as in residential and commercial building work.

## 2.6 The Capabilities and Limitations of Geomembranes

Because of its impermeability, a geomembrane offers many solutions to the challenge of expansive soils. Geomembranes reduce moisture change and therefore minimize volumetric changes in expansive soils. In a transportation facility such as a highway, street, railroad, or airport runway, reducing the volumetric change means lessening the distortion of the facility and maintaining the integrity of the original construction capabilities. The expense and inconvenience of repair work are decreased, and the life of the facility is extended. These benefits have been realized in building developments as well.

Experience has shown that the thinner geomembrane fabrics of 8 to 10 mils frequently have sufficient strength to avoid being torn by the passage of rubber-tired construction equipment. Durability of the geomembrane will be discussed in greater detail in Chapter 15. With appropriate care a geomembrane will last a considerable length of time. It is a very durable material.

It is not, however, without limitations. Degradation caused by ultraviolet rays, such as those of sunlight, can damage a geomembrane. Also, although geomembranes resist tearing from rubber-tired construction equipment, a heavier geomembrane is required if materials other than rubber are used on the wheels of similar vehicles. Furthermore, careless placement tends to diminish its barrier qualities that help minimize the destructiveness of the expansive soils.

The long-term durability of geomembranes has been a significant issue in their use. DeBerardino and Hawkins report uncovering a Typar placed 20 years earlier on a Delaware roadway. The geomembrane had been placed on a rural highway subgrade and covered with a base course. Some signs of wear were observed, but there was a deterioration of significant design characteristics. In a larger research project, M. A. Falk and G. M. Hager of the Wyoming Department of Transportation examined six geomembrane placements on their highways. In every case the geomembrane showed reduced strength, lower puncture, and Mullen Burst minimum roll test values of 140 psi. In each case they reported the roadway continued to have a smooth riding surface. The geomembrane-protected sections exhibited none of the distortions associated with expansive soil damages. A similar study by Rudolph Bonaparte offered further evidence of high geomembrane durability, which has alleviated concerns about their use.

The durabilities of PVC geomembranes have been discussed by Tisinger and Giroud [3]. A 9-year-old cap on a south Florida landfill using a 20-mil PVC geomembrane was found to have lost some of its plasticizer. Of the several samples taken, all but one were still soft. The PVC geomembranes 31 mils thick that were used in the construction of the Terzaghi Dam in Canada were found 14 years later to continue to perform as intended. The U.S. Bureau of Reclamation began using geomembranes in canal construction in the 1960s. After 19 years of service, they found the least aging in the 10-mil geomembranes with higher plasticizer percentages. (See Chapter 15 on durability for a discussion of these and other related cases.)

## 2.7 The Effectiveness of Geomembranes

There can be little as satisfying for an engineer, a builder, a developer, a banker, or a taxpayer than to see a facility continue to serve without the constant need for repair. Beyond the emotion of satisfaction, the freedoms from additional unexpected costs, litigation, and the negative image of incompetency are all benefits of well-done construction projects. A south Texas tale comes to mind: A district engineer informed a state legislator of the maintenance work, a considerable asphalt level up, that was going to be done on a stretch of interstate highway in her district. The legislator's response was, "You've been doing that for years." The district engineer quickly responded, "But this year we're going to use geomembranes vertically to cut out the bumps in the roadway surface." They used geomembranes, and for over a dozen years no maintenance work was required on that stretch of highway. The result was a saving of money and face and an enhancement of the image of the engineer and the state highway department.

In the United States there have been an increasing number of class-action suits as well as a rash of individual claims related to damages caused to homes and businesses by expansive soils. These claims have been filed in the Corpus Christi area of Texas along the coast of the Gulf of Mexico. Lawsuits and claims have also taken place in the San Antonio and Dallas areas of Texas. Claims and suits have been significant in the Denver, Colorado, area. This same pattern has occurred in California's Contra Costa County and in the Imperial Valley.

The significant number of projects where success has been reported using geomembranes to minimize moisture change in the expansive soils is an indication of their wide applicability. Though they have not been studied in as many building projects, the studies that have been made give favorable reports. The geomembrane can reduce the movements of expansive soils and thereby enhance the image of the engineer and builder as successful practitioners.

## 2.8 References

1. R. K. Frobel. "Geosynthetics Terminology." *Manmade Fiber and Textile Dictionary*. Celanese Corporation, Charlotte, NC.
2. Jacqueline Kroschwitz. *Encyclopedia of Polymer Science and Engineering*, 2d ed., Wiley, New York, 1990.
3. L. G. Tisinger and J. P. Giroud. *Geotechnics Fabric Report*, October 1993.
4. H. H. Tan. *Drainage Under Pavements*. E.I. Du Pont De Nemours Company, Wilmington, DE, 1975.
5. Stephen J. DeBerardino and W. M. Hawkins. "A Durability Study of Typar Twenty Years After Installation." "The Smyrna Road Project." Transportation Research Board, Jan. 1994, Washington, D.C.
6. M. A. Falk and G. M. Hager. "Performance of Impermeable Membranes to Treat Swelling Soils: The Wyoming Experiences." Wyoming Department of Transportation, Cheyenne, 1994.
7. R. M. Koerner and J. P. Welsh, *Construction and Geotechnical Engineering using Synthetic Fabrics*. Wiley, New York, 1980.

# 3
# Expansive Soils

Expansive soils come in a variety of forms and they may be composed of clays or shales or other minerals. All expansive soils share a common characteristic: They react with significant volumetric change when their moisture content changes. Expansive soils have been identified on all the worlds' nonpolar continents to date—perhaps in time they will be found there too.

## 3.1 Expansive Soil Terminology and Physical Properties

*Expansive soils, swelling clays, shrinking soils, expansive shales*—all are terms applied to similar soil conditions. These soil types may be one of a variety of clay minerals such as kaolinite, illite, or smectite. The smectite group includes montmorillonite and bentonite [1]. One publication that addresses the subject in a noteworthy effort is Nelson and Miller's *Expansive Soils, Problems and Practice in Foundation and Pavement Engineering* [2]. Another useful publication is the Federal Highway Administration's *Expansive Soils in Highway Subgrades* [3].

Most clays are formed by the mechanical disintegration and/or chemical decomposition of rock material. Examples of disintegration factors are running water, wind, freezing and thawing, oxidation, or hydration. The combined mechanical and chemical actions are frequently classified as weathering. When rock material decomposes or disintegrates into a clay in place, it's classified as residual. In Israel, mountain soils through the Galilee, Shomron, and Judea have typical brown clays that are residual. Their parent material is sedimentary, dolomite, limestone, marl, and calcareous shales [4].

Other clays are formed from igneous rocks such as basalt and other volcanic formations. Transported soils also are sources of expansive clays and shales. Their parent materials can be alluvial sediments and basalts. Still other clays are formed by combinations of residual and transported material. Fu Hua Chen observes that soil classification systems frequently define clay particles as having an effective diameter of 2 μm more or less [5]. But particle size alone does not define a clay. Rather, the mineralogical compositions are the most important factors in these fine-grain particles. In addition, the electric forces on their surface are more important than gravitational forces.

Clay minerals absorb anions and cations. The most common exchangeable cations are calcium, magnesium, sodium, potassium, and ammonia. Cations (positive ions) are more readily absorbed than anions (negative ions). The cation-ion exchange rate for different elements may vary. This exchange capacity is the charge, or the electrical attraction, per unit of mass. Montmorillonites are 10 times more active in absorbing cations than kaolinites.

Kassiff, Livneh, and Wiseman's book draws on studies by Grim in 1953 to examine the two basic clay structures, the pyramidal or tetrahedral and the octahedral [4]. The tetrahedral is made up of four oxygen atoms with a silicon atom at its center. The octahedral, which is known as *gibsite,* is made up of six hydroxyl ions with an aluminum ion or magnesium atoms between two layers of hydroxyls. Koalinite is made up of gibsite sheets joined with silica sheets. The bonds are tight, which makes the kaolinite relatively stable. Since water is unable to penetrate between the layers, there is little swelling when the kaolinite is subjected to wetting.

Montmorillonite reacts differently. Though it too has gibsite sheets, it may include atoms of aluminum, iron, magnesium, or any combination of them. Additionally, the silicon in the tetrahedron may interchange with the aluminum atom. These changes result in a negative charge on the clay mineral. Cations are attracted to these charges, and the bonds between the montmorillonite units are relatively weak. Thus, water is able to penetrate between the sheets, causing separation and swelling.

## 3.2 Geographical Distribution of Expansive Soils

Expansive soils are a worldwide challenge. They've been identified on all the world's continents except thus far the polar ones. From damages

# Expansive Soils

in Arabia to a king's palace, to damage reports in the United States, Canada, Mexico, to the South American continent, Europe, Asia, the Indian subcontinent to Israel, South Africa, and Australia, these soils have made their presence known. A review of papers presented at the Seventh International Conference on Expansive Soils gives a global understanding of the challenge these soils present. The Federal Highway Administration report RD 79-51 [6] identified 40 of the 50 states as having these expansive soils (Fig. 3-1).

Studies in South Africa show the occurrence of these soils on one end of that continent. Other studies in the Sudan in east-central Africa have indicated the presence of the expansive materials there. Across the narrow stretch of sea to the Arabian peninsula of Asia, the states of Oman and Qatar have reported on their efforts to deal with these swelling materials. In Saudi Arabia similar problems have arisen—a study indicated that the expansive soils have caused cracking of a regal palace there. Israel began reporting studies in the late 1940s of the many instances of swelling soils in their small country. In fact, because of the prevalence of these soils in Israel, one of the early International Meetings on Expansive Soils was held in Haifa.

India and Burma have both reported numerous cases of damages caused by the expansive soils. India hosted several international and regional meetings to address these challenges and to search for solu-

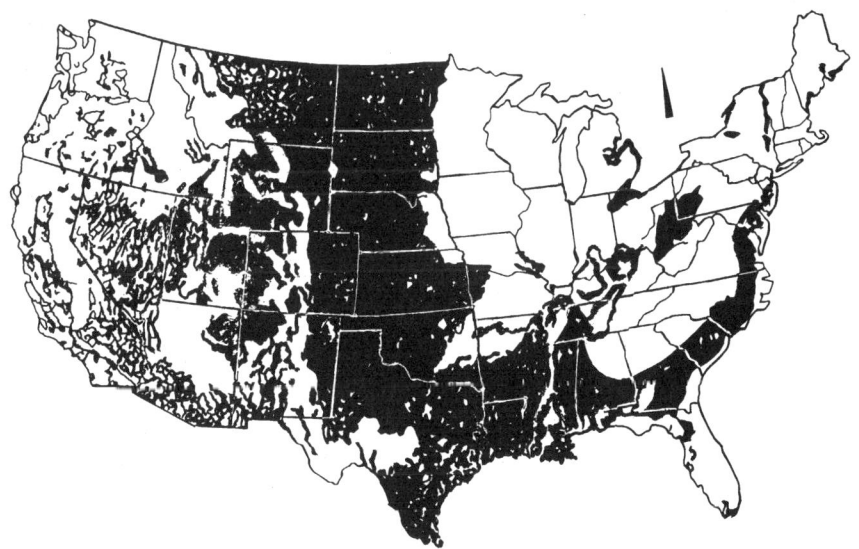

**Figure 3-1.** Expansive soils in the United States.

**Figure 3-2.** Expansive soils in Australia.

tions. Wiseman quotes Kitti in India who reported that 3 feet of lime-stabilized clay subgrade was more effective than 5 feet of gravel over the expansive soil. An early study by Wooltorton, indicated the extent of the damaging impacts of these soils in the Mandalay region of Burma [7] (Fig. 3-2). Further into Asia, the Chinese report that possibly 20 percent of their railway system, largely in the eastern part of the country, is built over expansive soils. To the south, on the subcontinent of Australia, these expansive soils are largely found along the southeastern parts of the country. Unfortunately, this is also the most populous part of Australia.

Moving around the globe to North America, Canada, the United States, and Mexico also face similar problems with the expansive soils. Reports from South America indicate the presence of these swelling clays in Brazil, Venezuela, Peru, and Argentina. The occurrence of these soils tends to follow the development of sophisticated built structures around the world. It is therefore likely that Russia and its one-time-aligned republics of Poland, Hungary, and Rumania will have to contend with these soils in the future in a creative manner.

## 3.3 Damages Caused by Expansive Soils

Studies in the 1930s identified two major human-made facilities upon which expansive soils have wreaked havoc. Wooltorton's investigations in the Mandalay district of Burma, where over 100 buildings suffered structural damages, found that the problems were not a failure in design. He further found that the damages were not due to faulty construction practices. Wooltorton's conclusion was that the expansive soils were to blame for the damages.

During this same period in the late 1930s and early 1940s, Porter, a research engineer for the Texas Highway Department, studied the problems faced on the highways in Navarro and Guadalupe counties. He correctly identified the movements in the concrete pavement as being caused by the swelling clays. These movements were monitored, and they were found to be related to the subgrades' expansive soils, a finding that was confirmed by the Atterberg limits. The material was reported in a meeting at Harvard in 1936. Willard Simpson, Sr., a San Antonio, Texas, consulting engineer, commented at the conference and indicated that he had 20 years of studies on expansive soils. Wiseman's paper notes that this conference was one of the early engineering gatherings in which the impacts of the expansive soils were examined.

Subsequent conferences and studies have brought to light that as more involved structures have been constructed and a wider range of building materials has been used, many more built facilities have been affected than previously. Pipelines and canals, airfields, and highways and streets in the United States and Israel have been identified as having to deal with these expansive soils. Damages have been recorded in the countries already noted as well as in South Africa, Canada, and Australia. China has reported substantial damages to their railroad system caused by these soils. Currently in the United States private home owners have become involved in a series of class-action suits. The claims are that the engineers, builders, and other entities knew of these soil problems. Since they did not adequately protect the purchaser, they are liable for damages. The most recent suit in Denver, Colorado, represents almost 1000 home owners.

## 3.4 Environmental Impacts of Expansive Soils

Sanitary landfills are an immediate concern in the environmental arena. Too many years of use-it-and-throw-it-away consumption patterns has led to a variety of refuse problems. In the United States the commonly used method of dealing with solid waste has been to haul it to an old pit somewhere away from town or to a nearby body of water and dump it there. As our population has increased, so has this challenge. It might have been appropriate for earlier inhabitants in a more nomadic situation to toss it as they moved, but generally land has become more valuable and the people living nearby raise questions about dumping practices. People are uncomfortable about sites chosen for solid waste disposal, even in fairly lightly populated areas such as far west Texas. Residents of this area have objected to New York waste being dumped near their homes, however lightly.

When the disposal site is over an expansive soil, significant environmental questions come into focus. Will the liquid seep through cracks in the subgrade and pollute a water supply? Substantial problems can arise even if the disposal site is not established over an expansive soil. The Love Canal, Hooker Chemical's waste disposal site, resulted in a series of expensive, environmentally damaging situations. When a waste site is located over an expansive soil, there is a great potential for creating severe health problems, polluting drinking water, causing genetic and birth defects in the yet unborn. In these cases, it is of some reassurance to see that geomembranes are being used in ever-increasing quantities to prevent such catastrophes. The European countries, particularly Germany, have addressed recycling with greater vigor than the United States. This recycling action has reduced the demands for waste site space and sanitary landfills. To a certain extent, recycling simplifies the challenge. However, the problem does remain to some extent, and expansive soils exacerbate it. Geomembranes help.

## 3.5 Financial Impact of Expansive Soil Damage

In the 1970s Jones and Holtz estimated that damages in the United States from expansive soils amounted to more than $2 billion annually [8]. Jones explained the basis of this survey. A variety of government agencies, contractors, builders, and developers were asked to estimate the damages caused by these soils annually. They were asked to give a low- and a high-figure estimate. Duplications were not included, and the low figure was the only estimate used. A decade or so later a new estimate was presented at the Fourth International Conference on Expansive Soils. The new figure was in the $9 billion range [9]. There is little doubt that these damages in the United States alone now well exceed $10 billion annually. This estimate was noted as exceeding all the damages wrought in the country by hurricanes, tornadoes, floods, and earthquakes.

## 3.6 Costs of Expansive Soils Disasters Compared with Other Natural Disasters

How do these dollars compare to damages caused by other disasters? In the fewest words possible, the expansive soils damages exceed all other natural disasters. This includes flooding: The 1995 Mississippi floods were estimated to have caused $10 billion in damages, less than the total damages for that year caused by expansive soils. It exceeds damages from hurricanes, tornadoes, snow, and ice storms. While these swelling soils have caused some increased attention lately compared to previous years' silence, there is still not enough interest in them.

## 3.7 Long-Term Benefits of Studying Expansive Soils

Over time there will be an increased need for substantial structures to house, educate, and care for the increasing numbers of people brought forth onto our earth. Greater awareness of the location of expansive soils will result from increased construction and maintenance projects around the world. With the use of more effective control measures, addressing these soils will reduce the demand for financial resources that are no longer seen to be inexhaustible.

As has taken place in several locations, the use of geomembranes has controlled or minimized damages from these soils. Their use saves maintenance money and reduces the anxiety that a vehicle traveling over a pavement severely distorted by the expansive soils might turn over. Their use may also reduce the anxiety that the house might collapse if those cracks from the expansive soils enlarge. Furthermore, in providing a solution to the challenge of expansive soils, the use of geomembranes will enhance the image of the civil engineer, the builder, the contractor, and the developer. Using geomembranes may also reduce the likelihood of costly class-action liability suits.

As our body of knowledge about expansive soils grows, so will our ability to prevent damage from them. Presently, we know that expansive soils could be clays or shales. Their identification is not difficult. Atterberg limits, suction values, Thornthwaite indices, potential vertical rise tests, and calculations all provide aids in their identification. These soils tend to dry and crack during periods of little rain. When exposed to rainfall or to cracked water-carrying pipes, they absorb water, which results in considerable volume change and swelling.

In the United States a report by the Geotechnical Laboratory of the U.S. Army Corps of Engineers, Waterways Experiment Station, for the Federal Highway Administration (FHWA-RD-79-51) shows the presence of expansive soils in 40 of the 50 states. Activity and the amount of damages caused by these soils vary. The multiyear COE study had a five-member Expansive Soil Technical Advisory Group (ESTAG), which included this author. All the members of the ESTAG had previously written about on-the-job practical experiences in which expansive soils were part of the construction challenge. This sharing of what worked and what didn't provided information that could be used in other construction projects. Engineers serving on the ESTAG came from Arizona, Colorado, Mississippi, South Dakota, and Texas. All had construction and design experience. All learned from other participants different ways of dealing with expansive soils, an experience without price for all.

In 1973 Jones and Holtz reported in ASCE's *Civil Engineering* magazine that these expansive soils damages amount to $2.3 billion annually. They arrived at this figure by surveying federal, state, and local agencies, and contractors and developers and asking for their high and low estimates of damages caused by these soils. They took the low figure for their composite total. In their 1981 report, the U.S. Department of Housing and Urban Development estimated shrink-swell damages at $9 billion annually [8]. The shrink phenomena is frequently closely associated with the swelling characteristics of expansive soils. It certainly would be safe to estimate damages caused by expansive soils in the United States to exceed $10 billion annually. These expansive soils can be seen to cause more damages in this country annually than any other natural hazard including earthquakes and floods [2].

The damages caused by expansive soils amount to a national disaster occurring annually. Recently home owner suits relating to these damages have been filed in Corpus Christi, Texas, on the Gulf coast, Denver, Colorado, in California's Contra Costa County in the bay area, and in the Imperial Valley area near San Diego. The Denver occurrences are not new. They've been reported for over three decades by Fu Hua Chen [5]. Chen further identifies other areas around the world in which heaving often occurs, citing a study by G. W. Donaldson. Of particular note is that South America is reporting heaving in its northeast corner, in Colombia and Venezuela, and in its southeast area, in Brazil. In Africa the reported areas are the northeast segment including Morocco, the west-central countries of Sudan and Ethiopia, and the southern tip area of South Africa.

In the United States Chen credits the U.S. Bureau of Reclamation as the first to recognize the swelling soil problem in 1938. They addressed it in their design of a steel siphon for the Owyhee project in Oregon. Since that time, engineers in the United States and abroad have studied expansive soils with increasing intensity. The first national meeting of significance on the subject of expansive clays was held at the Colorado School of Mines in 1959. The first International Research and Engineering Conference on Expansive Soils was held at Texas A&M University in 1965. It was also the site of the second conference held in 1969, while the location of the third was Haifa, Israel, in 1973. The total has now reached seven conferences. In the summer of 1995, the University of Colorado at Denver held its second annual symposium on expansive soils. The awareness grows, and the interest mounts.

Where Chen focuses on the expansive soils' impact on buildings, *Pavements on Expansive Clays* by Kassiff, Livneh, and Wiseman looks at a transportation facility [4]. Jones and Holtz's study has indicated that these facilities have suffered half the damages caused by the expansive soils [8]. Israel has many expansive soil problem areas, from the Galilee in the north to the edge of the Negev area, and with the exception of the sands of the coastal plain and the Judean hills to the east, clays abound. They distort roadways, airport pavements, buildings, and canals. Australia has reported through the Victoria Roads (VICRoads) office a profusion of expansive soils in their southeastern to south-central areas where the majority of their population lives. It creates a particular problem affecting roads and buildings. China has reported similar problems with expansive soils. Up to 20 percent of their railway system is estimated to be built over expansive clay subgrades. As they have in rail systems everywhere else in the world, destructive soil movements have taken place. Maintaining the rail system requires a substantial annual expenditure. An investigation of building failures in the Mandalay area of Burma found that the damages were not due to shoddy construction, as had first been assumed. Instead, the culprit was found to be expansive soils.

South Africa has also been impacted by these expansive soils. Reports of the destructiveness of these soils on their highway and railway systems as well as to their buildings again indicate the widespread impact and presence of expansive soils. A paper by Blight and colleagues describes the damages created by the expansible soils in South Africa and the efforts to control them [10]. Blight's study reports damage to private residences, an apartment building, and shopping complexes.

Blight indicates that damages might have been avoided by preheaving the expansive soils.

Similar reports from buildings being negatively impacted by expansive soils in the Sudan provide further data on their widespread presence. Both Wooltorton and Krazinski have reported on the damages these soils create there.

Addressing the challenge of expansive soils will help in preventing the destruction they cause. Doing so will also provide the opportunity to save money in both maintenance and legal costs.

## 3.8 References

1. W. K. Wray. *So Your House Is Built on Expansive Soils.* American Society of Civil Engineers, New York, 1995.
2. John Nelson and D. H. Miller. *Expansive Soils, Problems and Practice in Foundation and Pavement Engineering.* Wiley, New York, 1992.
3. Federal Highway Administration. *Expansive Soils in Highway Subgrades.* U.S. Department of Transportation, Federal Highway Administration, Washington, D.C., 1980.
4. G. Kassiff, M. Livneh, and G. Wiseman. *Pavements on Expansive Clays.* Jerusalem Academic Press, 1969.
5. Hua Fu Chen. *Foundations on Expansive Soils,* 2d ed. Elsevier, Amsterdam, 1975.
6. D. Snethen. *Technical Guidelines for Expansive Soils in Highway Subgrades.* U.S. Army Corps of Engineers Waterways Experiment Station, Federal Highway Administration. Washington, D.C., 1979.
7. Wiseman. *Experience with Swelling and Collapsing Soils.* Monash University, Melbourne, Australia, 1994.
8. D. E. Jones and W. G. Holtz. "Expansive Soils—The Hidden Disaster." *Journal of Civil Engineering Proceedings of the American Society of Civil Engineers,* vol. 43, August 1973.
9. J. P. Krohn and J. E. Slosson. Assessment of Expansive Soils in the United States. *Fourth International Conference on Expansive Soils,* vol. I. Denver, 1980.
10. G. E. Blight, K. Schwartz, H. Weber, and B. L. Wild. "Preheating of expansive Soils by Flooding—Failures and Successes." Seventh International Conference on Expansive Soils. Dallas, 1992.

# 4
# Responses, Studies, and Theories

You will have the chance in this chapter to get a further look at what has happened to date in search of understanding these expansive soils. They are attracting more attention not just from owners and their attorneys but from engineering professionals like yourself and the people you work with in construction. The material presented will increase your feeling of a partnership with many distinguished professionals around the world who are facing this expansive soils challenge. Ultimately it's a question of how to build on this material with success and profit.

## 4.1 The Atterberg Tests

Atterberg's soil analysis methods were the first successful qualitative tests for categorizing clays [1]. As described by Terzaghi, the Atterberg laboratory tests involve the determination of a series of limits: liquid, plastic, and shrinkage [2]. The numerical difference between the liquid limit and the plastic limit is the plasticity index of the soil. The difference between the soil properties as determined by the Atterberg tests—types of clays and adsorbed ions—are noted in detail by Chen. The greater the cation-exchange capacity of the clay, the greater the effect of changing the adsorbed cations. To chemically stabilize an expansive soil, the ionic concentration of the free water and base exchange phenomena must be increased. In later decades the plasticity indices

Atterberg developed became key indicators and sometimes sole factors in identifying swelling clays [3].

## 4.2 Continuing Studies by Universities

Universities have frequently provided the setting for soil studies. *Elementary Soil Fundamentals* credits Dr. Karl Terzaghi as introducing engineering approaches into the study of soils in 1923 [4]. Terzaghi is generally recognized as the Father of Soil Mechanics. In their book, Lambe and Whitman summarize highlights of Terzaghi's career [5]. In the early days of his professional life, Terzaghi attempted to find rational approaches to earthwork engineering problems. These efforts culminated in 1925 with the publication of his book *Soil Mechanics*. Between 1925 and 1929, Terzaghi was at MIT establishing the first U.S. program in soil mechanics. It emphasized the intimate relationship between civil engineering and soil mechanics, which he viewed as an integral part of the profession.

In 1928 Terzaghi became a professor at Harvard University. There he developed and presented his course in engineering geology. Through 1960 Terzaghi published over 250 papers, reports, and books. He received honorary doctorates from nine countries, won the American Society of Civil Engineers' Norman Prize four times, and was the first president of the International Society of Soil Mechanics and Foundation Engineers. Terzaghi died in 1963, and the ASCE in commemoration of his monumental work, established the Terzaghi Lecture and the Terzaghi Award [6].

Gillette credits Thomas H. MacDonald, chief of the U.S. Bureau of Public Roads (BPR), with making possible further soil science studies. Gillette's 60-page booklet, which sold for 65 cents in 1936, presents extensive use of Atterberg's tests in soil and other significant identification guidelines. Grain size, cohesion, and friction factors are offered for analysis. Measures of capillarity, permeability, elasticity, and internal friction enabled the readers to assess differences among sands, clays, silts, and colloids. MacDonald established a soils laboratory at the Bureau of Public Roads. Terzaghi worked closely with this effort, later retiring from Harvard and relocating at the University of Illinois. MacDonald, on retiring from the bureau, became head of the Texas Transportation Institute at Texas A&M University.

A bulletin of the Agricultural and Mechanical College of Texas, now Texas A&M University, authored by Henry Porter, research engineer, Texas Highway Department, entitled *Roadway and Runway Soil Mechanics*

*Data,* was published in 1942. The bulletin includes a summary of Parts 1 through 11 of studies that concentrate largely on clay soils. The bulletin, Parts 10 and 11, examines in detail the density, moisture content, strength of consolidated clays, and moisture of clays underneath pavements. Porter states increased loadings of pavements at highways and airports make durability of the substructure, the subgrade soils, more important. Testing of moisture and densities were conducted and reported in detail. Sections of highways in Guadalupe and Navarro counties were studied relating moisture to density and movement (Fig. 4-1).

Porter's studies in 1933 emphasize the importance of determining the most economical method of providing a subgrade contributing to a smooth riding pavement. The summary notes that the strength of a sand is greatest when the sand is wet and weakest when dry. With clays, the variation in moisture content is reflected in the load-bearing capacities. Considerable data related to clays, moisture contents, and bearing capacities are included along with the importance of maintaining moisture contents. Ponding of a highway subgrade was also reported by Porter. In his acknowledgments, the names of Gillette, whose pamphlet was previously discussed, of John Focht, Sr., professor of highway engineering, University of Texas, father of an ASCE national president, and Chester McDowell whose work will receive further attention, are among those mentioned.

The University of Texas Center for Highway Research published the 118 series of reports dealing with expansive clays. The *Theory of Moisture Movement in Expansive Clays,* 118-1, was the first, authored by Robert L. Lytton, now professor of civil engineering at Texas A&M University, and it was published in 1969. The report presents a theoretical discussion of the moisture movement in clay soils. Lytton states that the understanding of the computer programs to be developed in subsequent studies will lead to practical applications. The report also examines permeability, soil suction, and unsaturated soil stress. The movement of moisture in clay is compared to the flow of current in electricity. Soil analogs of electrical conductance, voltage and "something like inductance" are described. Suction is normally higher when the soil is dry and lower when it is wet. The report identifies a half dozen factors that influence suction.

The University of Texas CHR Report 118-2, *Continuum Theory of Moisture Movement and Swell in Expansive Clays,* authored by D. R. Nachlinger and R. L. Lytton, was published in September 1969. This theoretical study views expansive clay as a macroscopically continuous material. It uses a mathematical mixture theory of continuum mechanics. In the abstract it is seen as providing the foundation for understanding the mechanism of expansive clays. The conclusion observes

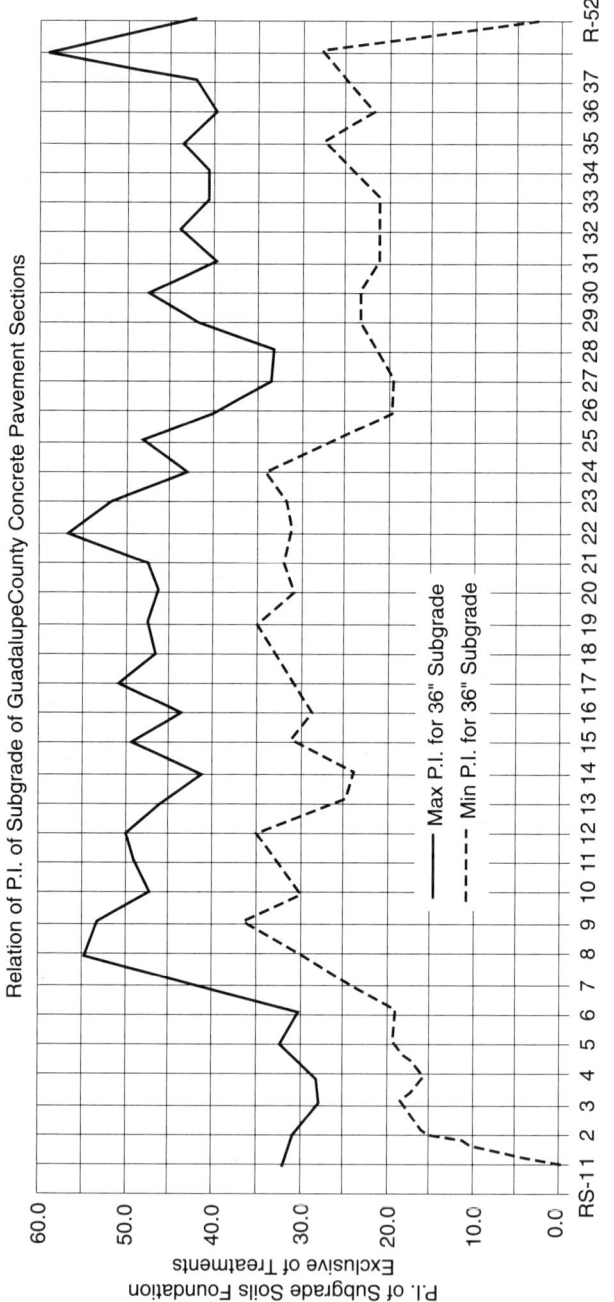

**Figure 4-1.** Plasticity indices in Guadalupe County, Texas.

that the basic development of balance laws are known, which provide the constitutive equations and the interpretation of the various results.

The reports 118-3 through 118-8 continue the study of the expansive clays. Three examine the reactions of the clay to a ponded cut. Gordon Watt, now deceased, coauthored the first with this author who was the sole author of the second. Through the pre- and postconstruction testing of the U.S. 90 freeway in southwest San Antonio, the authors found a clear zone of activity where the significant moisture changes take place. Results showed a rise in subgrade elevations as a result of the ponding. The question of whether the ponding could preswell the subgrade was answered, and this method was found to be successful. The consideration of the zone of activity of moisture change would prove useful when considering the question of how deep to place the geomembrane vertically [7 and 8].

Chester McDowell coauthored a third ponding study in this series, reporting on the long-term impacts of the project on IH 35 in the Waco area. The ponded sections required less maintenance work than the ones not ponded, which was evident in the pavement cracking and heaving [9]. Similar affirmative responses were found by Paul Teng, then Mississippi DOT research engineer. A ponded project in the Jackson area had holes drilled 20 feet deep, which were filled with pervious material and then ponded. It was reported to be one of the few in the area not heaving or experiencing swelling. South Africa reported similar success with ponding in their tests also.

The University of Texas research report *An Examination of Expansive Clay Problems in Texas* included a test site on IH 30 [10]. It was in a 28-foot clay-cut section 10 miles northwest of New Boston, in Bowie County, northeast Texas. Among its conclusions was that nuclear density and moisture readings were suitable for determining these values in field conditions. The report also observed that major moisture changes took place in the top 6 feet of the subgrade of the New Boston site.

## 4.3 Texas Department of Transportation Studies

The Texas Highway Department has gone through several evolutions and name changes. The THD first became the State Department of Highways and Public Transportation and then later became the Texas Department of Transportation, or TexDOT. Throughout its history, however, it has not retreated from its commitment to the research and study of the challenges of expansive clays. It is noteworthy that, in his reports, Porter is identified as "Research Engineer, Texas Highway Department." Presently TexDOT contracts most of its research to state universities in particular areas. The department also maintains area committees that focus on a variety of transportation challenges. One of these committees focuses on pavement systems, which includes subgrade studies. These studies are frequently directed toward examining methods of meeting the swelling soil challenge. The TexDOT committee prioritizes the matters brought before it by representatives from the districts and the Austin Division offices. The studies are conducted by the universities with a department research contact person. The University of Texas at Austin, El Paso, San Antonio, and Arlington have all been involved, as has the Texas A&M University. Currently one of the major projects being evaluated and reported is the use of geomembranes as vertical moisture barriers in the widening of IH 45 near Waxahachie.

TexDOT is not the only state transportation agency studying and publishing reports. Not all have addressed the expansive clay challenge and the use of geomembranes, but they have prepared considerable reports and published studies. States neighboring Texas have been particularly active in this area. The Oklahoma DOT has published several reports on the lime and cement treatment of expansive clays. Under James Nevels, they have worked on and published reports about the use of suction on their highways. The New Mexico DOT has similarly conducted studies and issued reports on their work with swelling soils. The Arizona DOT has been very active in this area also and has used geomembranes as moisture barriers on a number of its highways.

The CalTrans, Kentucky, Wyoming, Mississippi, South Dakota, Colorado, Montana, Utah, and Florida DOTs have reported on their efforts to control expansive soils, in some cases using geomembranes as moisture barriers. In the 1970s the Colorado Highway Department used a variety of techniques in efforts to control these subgrades. They published the results, and their use of heavy asphalt applications on a swelling subgrade led to the first geomembrane use on a state system highway. The South Dakota Highway Department published Eugene McDonald's report on the use of a geomembrane as a vertical moisture barrier on a major highway. Some of the early Mississippi reports looked at the use of ponding that involved drilling holes and filling them with pervious materials. A more recent report has dealt with the use of geomembranes placed horizontally on the expansive subgrade.

## 4.4 Federal Highway Administration Studies

Before there was a Federal Highway Administration (FHWA), there was a Bureau of Public Roads. Under the able direction of Thomas

MacDonald, referred to earlier in this chapter, many powerful steps were taken to develop this nation's highways. One was the BPR's efforts to assist Terzaghi in the understanding of the challenge presented by the subgrade beneath the nation's highways. The BPR's awareness of expansive soils led to the funding, decades later, by the FHWA of a substantial research project at the U.S. Army Corps of Engineers, Waterways Experiment Station (WES). The initial report was entitled *A Review of Engineering Experiences With Expansive Soils in Highway Subgrades*. The authors were Donald Snethen, Frank Townsend, Lawrence Johnson, David Patrick, and Philip Vedros. Published in 1975, the report noted that the impact of damages caused by the expansive soils to the nation's streets and highways was $1.1 billion annually. Due to their prevalence throughout the United States, it is virtually impossible to avoid expansive soils in the construction of the nation's streets and highways. The interim report includes an extensive literary search. It provides outstanding maps of different sections of the United States indicating areas where expansive soils have been identified. Mineralogy and sampling techniques, along with a review of preconstruction methods, to deal with these soils are explained. The University of Wyoming report on efforts to treat the subgrades on IH 80 west of Laramie is summarized. The various alcohol, formaldehyde, and other chemical conditions were tried as corrective measures. The conclusion was that none proved effective. Lime treatments—shallow, deep, and slurry—are reviewed, and a wide variety of other methods are examined. The report explains the importance of maintaining the proper moisture density ratios.

The WES/FHWA project's final report was issued in June 1979. Authored by Dr. Donald Snethen, *Technical Guidelines For Expansive Soils In Highway Subgrades* noted that damages to U.S. streets and highways caused by the expansive soils had increased to $1.7 billion annually. Minimization of the damages is the goal of the 4-year study's recommendations. Guidelines offered include the locations of expansive soils areas. The report provides other sources of information, field exploration and test methods, and classification systems using index and soil suction properties with treatment alternatives. The report also gives predictions of volume changes that occur with different treatment modes. Some of the participants in this study were the members of the Expansive Soil Technical Advisory Group (ESTAG). These five engineers had worked, and written, in the field on construction projects where expansive soils were involved. As an ESTAG member, the author learned from the others, which led to the successful use of geomembranes in Texas. Other publications and presentations by the FHWA followed their 1979 report. Some focused closely on pavement design, geotextiles, and expansive soils. The information is useful, but measures to control the challenge of expansive soils are not examined thoroughly.

## 4.5 Transportation Research Board Meetings and Publications

A significant group aiding the effort of better understanding the expansive soil challenge and its possible solution is the Transportation Research Board (TRB) of the National Academy of Science. Initially the TRB Committee on Environmental Factors Other Than Frost, A2L06, focused on the questions posed by the expansive soils. Meetings of the committee took place over several decades during the annual meeting period. Papers were presented and published by committee members who came from the United States and other countries. Presentations from abroad included those by engineers from South Africa, Egypt, Saudi Arabia, Canada, China, and Israel. This sharing of knowledge and experiences is what TRB is all about. These conferences have helped engineers from the United States and other countries cope with the challenges of the expansive soils.

As the use of engineering fabrics became more widespread, a TRB committee was formed to examine them. This committee, A2K07, has sponsored several sessions and publications on the use of these fabrics. Though some have addressed the use of geomembranes in the control of expansive soils and its durability, the reports have not played a major role in focusing on the applicability of geomembranes in controlling the destructive movement of clays and shales.

## 4.6 Suction Testing

Terzaghi and Peck's book *Soil Mechanics in Engineering Practice* [2] touches on the impact of suction in the study of expansive soils. The question of suction is covered in greater depth in Kassiff, Livneh, and Wiseman's book. The total potential of soil water consists of gravitational potential, osmotic potential, and suction. *Suction* results from a decrease in potential negative pressure caused by the interaction between fluid and solid—that is, by capillary forces.

Kassiff, Livneh, and Wiseman further note that soil situated above groundwater level will have osmotic potential and suction. Suction results in a decrease in the pressure head with an increase in elevation above groundwater level. Soil that is in a condition of moisture equilibrium shows a linear decrease in pore pressure. Water is held in porous materials by surface tension and adsorption forces. Clay soils contain particles whose surfaces are negatively charged. Forces of absorption exist in addition to surface tension and are affected by the salts in the soil. A decrease in the pressure head in the soil water above the groundwater level is also known as *suction*. This terminology is interchanged with *capillary potential,* as used in Buckingham's work in 1905. There the soil was assumed to behave as a system of capillaries. Suction is an important factor controlling the effective stress of the soil. Its changes in soil lead to movement of moisture from low to high suction. Suction is also an important factor controlling effective stress in the soil. Croney et al. and Black are credited with finding a clear relationship between suction and the California bearing ratio values.

Because of the wide range of suction values, a logarithmic scale is used for describing them. It's shown as pF, and G. Kassiff, M. Livneh, G. Wiseman, and colleagues attribute its introduction to Schofield in 1935. He defined suction in terms of the logarithm of the negative pressure expressed in centimeters (cm) of water. Soil suction can be measured with a tensiometer. Other direct measuring can be done with a suction plate or a porous membrane or by consolidation. Indirect methods call for the use of freezing point depression, vacuum desiccator, sorption balance, and electric resistance gauges.

Measurement of suction in the field after construction is important from theoretical and practical standpoints. Continuing measurement will provide an accurate assessment and prediction of pavement movement and behavior. In the case of a highly compressible soil composed of a highly plastic clay, a large increase in suction accompanies a negligible change in moisture content, and as soil density increases, so does suction. Changes of the moisture content in clays are accompanied by changes in volume and pore geometry. As the clays dry out, suction increases, the soil shrinks, and the density increases. Water loss leads to a decrease in the clay volume starting from equilibrium with free water until the clay reaches the shrinkage limit.

Other factors influence the suction of compressible materials such as clays. These include geological history and the degree of disturbance of the clay. The relationship between suction and moisture content is considered to be influenced by the maximum exposure to moisture content during its geological history. The work of Croney and colleagues has demonstrated that an increase in temperature leads to a decrease in suction.

Chen emphasizes the suction impacts on clays [3]. Suction values vary widely. Using the logarithmic scale, 2 pF would equal 100 cm of hydrostatic head (205 lb/ft$^2$) while 4 pF would reach 10,000 cm (20,500 lb/ft$^2$). Chen states that if there are no moisture changes in the clay, there will be no volumetric change or movement. If a structure with a clay subgrade is protected from moisture changes and the moisture doesn't change, there will be no clay movements to damage the structure.

R. G. McKeen's work at the University of New Mexico Engineering Research Institute carried the suction studies forward with the development of the filter paper test. In the paper he coauthored with Thompson, he demonstrates the use of the filter paper technique in remedial efforts [12]. The same publication has James Nevels' article "The Use of Soil Suction in Analysis of Pavement Cracking" [12].

## 4.7 Texas A&M University Studies

The work of Henry Porter in the 1930s and 1940s has been mentioned earlier in this chapter. Porter identified the impact of plasticity indices and moisture contents on highways in Navarro and Guadalupe counties. They included the heaving of the highway between San Antonio and Seguin in south-central Texas and its relationship to the expansive clays. His papers were published by the Agricultural and Mechanical College of Texas, which led to an awareness of the challenge and an ongoing effort to address it.

Spencer Buchanan, a Texas A&M professor, carried the expansive soils study further. He later founded Buchanan and Associates, a consulting engineering firm specializing in soils and foundations. He maintained an active role in the U.S. Army Corps of Engineers, eventually rising to the rank of general. In a paper he presented at the Fourth International Conference on Expansive Soils held in Denver, Colorado, in June 1980, Buchanan's review of the history of the earlier meetings gives a record of his and A&M's commitment and involvement. The paper covers the work of many others and notes interest developing in the expansive clays in the early years of this century. Concerned that the expansive soils' movements were affecting structures, Willard Simpson, Sr., a consulting engineer in San Antonio, designed the drilled-shaft foundation. Buchanan reviews the development during and after World War II of lime treatments to expansive clays. This method of clay stabilization became widely used in Texas and in many of its neighboring states. Similar efforts were made in Africa, Australia, and Israel as a result of the growing awareness of the problem of expansive clays that was developing around the world. Buchanan particularly commends the work of Dr. Jeremiah Jennings in South Africa. In honor of Jennings' contributions, the Ftourth International Conference was dedicated to him.

The need to exchange the theoretical approaches most commonly used in Europe and the United States led to the First International Conference on Research and Engineering on Expansive Soils, which took place at the Texas A&M University in 1965. The second conference took place there in 1969, while the third followed in Haifa, Israel, in 1973. Several developments were presented at the third meeting, including one involving the use of a vertical moisture barriers in protecting residences. The use of posttensioned slabs to minimize damages from expansive clays was described.

Robert L. Lytton, professor of civil engineering at Texas A&M, followed in the tradition of working with the expansive soil challenge. His work has led to the publication of a series of research reports (1165) that address the possibility of using a vertical moisture barrier, primarily a geomembrane, to control expansive soils. Impacts and depths of placement, temperatures, root depths, crack and soil densities, and rainfall zones are all important factors affecting the success of geomembrane use. Between 1976 and 1993 about one dozen publications were presented and published on this subject.

In the *Effectiveness of Controlling Pavement Roughness Due to Expansive Clays With Vertical Moisture Barriers* [13], the authors conclude that geomembranes used as vertical moisture barriers are effective with certain limitations in minimizing destructive soil movements in highways. This generalization includes only the effectiveness of geomembranes in a medium-cracked soil. Even in these soils, the geomembrane is viewed as ineffective in extremely dry climates or in ponded drainage conditions or in tightly cracked soils. Another conclusion reached by the researchers was that when a vertical moisture barrier is placed, its depth should be at least equal to the surrounding root depth. The study is based on data collected at 6 Texas sites from the northeast corner to the far west of the state. The report further concludes that substantial site soil investigation should take place prior to any design decisions. It also commends the suction compression index as a powerful tool because it is a measure that can be determined easily with the use of available charts. A FLODEF computer program has been developed that can predict suction changes and vertical movement of expansive soils. Although more than 25 transportation projects in Texas had used geomembranes as moisture barriers at the time the report was written, the report's conclusions and recommendations are limited to the research conducted at only six of them.

## 4.8 Ongoing International

Delwyn Fredlund at the University of Saskatchewan in western Canada continues suction studies that have culminated in a recent publication

[14]. This report examines studies beginning in the 1980s and is titled *State of the Art Development In The Measurement of Soil Suction* [15]. Richard Bathurst at the Royal Canadian Military Academy has done and continues to do investigative studies that involve the use of engineering fabrics in controlling expansive soils. Leonard's investigations have been ongoing for decades at the University of Ontario.

Australian studies of expansive soils have been prolonged and continuing. Gordon Aitchison, chief, Office of Scientific and Industrial Research, did much of the pioneering work in soil mechanics in Australia. In other research, Thornthwaite developed an index to describe the relationship of transevaporation to swelling soil. The Thornthwaite index has gained worldwide prominence and usage. James Holden,* with the Victoria Roads (VICRoads) research center, is conducting long-term studies of the use of geomembranes in both road and building construction. These studies have already led to considerable improvements in geomembrane use.

Swinburne University, working cooperatively with VICRoads, has developed a program that is actively seeking solutions to the problem of expansive soils. The participants in this program, directed by the deputy head of civil engineering, Kerry McManus, work closely with Holden, VICRoads, and the Australian Road Research Board (ARRB). This cooperative effort has led to the development of the slim-line trencher. This trencher excavates to a depth of 6.5 feet, which is about the depth of the zone of activity in Victoria. The trench it cuts is about 3 inches wide. After vertical placement of the geomembrane, flowable backfill equipment, also developed in this program, is used.

Currently several related projects are continuing. A study is being conducted by a graduate student, now lecturer, Robert Evans, on the ARRB laser-based profilometer to measure movements at the Doonen highway test site. Plans call for studying the use of soil suction inside and outside the barrier to measure its effectiveness in stabilizing the subgrade. Concurrently Anthony Mann, working with Holden and McManus at Swinburne University, is developing a more cost effective method for vertical moisture barrier placement. The power spectral density (PSD) is the basis for this method, and it evolved from the raw profile data received by VICRoads. The PSD method is being used with the plan to reduce future maintenance costs. The cooperative work by these three Australian groups is making significant contributions to the existing body of knowledge on expansive soils.

---

*Now retired from VICRoads. Presently with Swinburne University.

South Africa, similarly faced with expansive soils problems, has addressed them through research. Much work has been done at the University of Witwatersrand in this field. Professor Jeremiah E. B. Jennings, the KANTHAK Professor of Civil Engineering at the university, was one of the founding members of the International Society of Soil Mechanics and Foundation Engineers. Kassiff, Livneh, and Wiseman's book notes the contributions of A. A. B. Williams, N. E. Simons, De Wet, and Geoffrey E. Blight at Witwatersrand University in seeking solutions to control these destructive expansive movements.

Substantial work on these soils began at Israel's Institute of Technology (The Technion), led by Joseph G. Zeitlin, professor of civil engineering, in the 1950s. Over time he was joined by Amos Komornik, G. Kassiff, Moshe Livneh, G. Dalyah, Wiseman, S. Frydman, and Jacob Uzan in this work. Israel's rapid growth has led to an increased use of geomembranes.

A great deal of experimentation and research on expansive soils has created a substantial body of knowledge upon which engineers may rely to prevent disasters. In using geomembranes, you will not be plowing new ground.

## 4.9 References

1. A. Atterberg. *On the Investigation of the Physical Properties of Soils and the Plasticity of Clays.* Int. Mitt. Boenkunde, 1911.
2. K. Terzaghi and R. B. Peck. *Soil Mechanics in Engineering Practice,* 2d ed. Wiley, New York, 1967.
3. Fu Hua Chen. *Foundations on Expansive Soils.* Elsevier, Amsterdam, 1975.
4. H. S. Gillette. *Elementary Soil Fundamentals.* University of Oklahoma Press, Norman, 1936.
5. T. W. Lambe and R. W. Whitman. *Soil Mechanics.* Wiley, New York, 1969.
6. T. W. Lambe and R. W. Whitman. *From Theory to Practice in Soil Mechanics.* Wiley, New York, 1960.
7. Gordon Watt and M. L. Steinberg. *Measurements of a Swelling Clay in a Ponded Cut.* Research Report 118-6. Center for Highway Research, University of Texas, Austin, 1972.
8. M. L. Steinberg. *Continuing Measurements of a Swelling Clay in a Ponded Cut.* Research Report 118-8. Center for Highway Research, University of Texas, Austin, 1973.
9. R. L. McKinney, J. E. Kelly, and C. McDowell. *The Waco Ponding Project.* Research Report 118-7. Center for Highway Research, University of Texas, Austin, 1974.

10. J. R. Wise and W. R. Hudson. *An Examination of Expansive Clay Problems in Texas.* Research Report 118-5. Center for Highway Research, University of Texas, Austin, 1971.

11. G. Kassiff, M. Livneh, and G. Wiseman. *Pavements on Expansive Clays.* Jerusalem Academic Press, 1969.

12. D. Snethen, F. Townsend, L. Johnson, D. Patrick and D. Vedros. "A Review of Engineering Experiences with Expansive Soils..." U.S. Army, COE, WES, Vicksburg MS, 1975.

13. R. G. McKeen and R. Thompson. "Heave Prediction Using Soil Suction, A Case History." In *Soil Suction Applications in Geotechnical Engineering Practice,* edited by Wray and S. Houston. ASCE Special Publication No. 48, 1995.

14. R. Javatilaka, R. Gay, R. L. Lytton, and W. K. Wray. *Effectiveness of Controlling Pavement Roughness Due to Expansive Clays With Vertical Moisture Barriers.* Research Report 1165-2F, FHWA/Tx—92/1165—2F. Texas Transportation Institute, Texas Department of Transportation, and the Federal Highway Administration.

15. D. G. Fredlund and H. Rehardjo. *Soil Mechanics for Unsaturated Soils.* Wiley, New York, 1993.

16. D. G. Fredlund and H. Rehardio. *State of the Art Development in the Measurement of Soil Suction.* International Conference on Engineering Problems on Regional Soils, Beijing, China, 1988.

# 5
# Facilities Affected

## 5.1 Types of Facilities

A wide range of facilities are affected by expansive soils, and, with the exception of the polar continents, no country is immune from the hazard they create. As population increases, and with it the number of built structures, the dimensions of the expansive soil problem will also grow. Moreover, almost all types of built structures are vulnerable to expansive soil damage, even dirt roads and streets, wood-frame buildings, unpaved waterways, and wooden pipelines.

The costs of repairing expansive soil damages are also increasing. Since 1973, when the estimate of damages, in the United States alone was $2.3 billion, the costs have grown year in and year out [1]. The 1973 estimate was based on a survey seeking high and low estimates of damages caused by these soils. Then, only the low-estimate figures were added to reach the $2.3 billion. By 1980 the estimate of these damages was placed at $7 billion in the United States annually. Today the damages from the expansive soils in the United States exceed $10 billion annually, which is more than all the damages due to tornadoes, hurricanes, floods, and earthquakes in this country each year.

Observing the countries of the participating engineers at the 1980 International Conference on Expansive Soils gives evidence of the global impact of these soils. Engineers from the United States, Canada, South Africa, Australia, and Israel were in attendance along with engineers from Mexico, Venezuela, Chile, Peru, Argentina, Brazil, China, Greece, India, Romania, Russia, Poland, France, England, Spain, and Japan. The Seventh International Conference had representatives from the preceding countries plus Bosnia (formerly Yugoslavia), Bangladesh, Hungary, Saudi Arabia, the Oman, and Thailand.

The facilities affected by the expansive soils range from an Arabian king's palace to homes in South Africa, Australia, Israel, and the United States to railroads in China, France, and Russia. The Arabian peninsula has many soils classified as "sands" but some are actually expansive clays. The problem is so severe that swelling soils have caused considerable cracking in one of the Saudi king's palaces. Expansive soils also exist in Saudi Arabia's neighbors to the southeast, Oman and the Gulf states, and in Jordan and Israel to the north. Israel has suffered damages from these soils to their small country's roads, buildings, canals, and airfields. On the southern tip of the African continent, swelling soils have similarly damaged residential buildings in South Africa as well as their highways and rail lines. In southeast Africa, Sudan has had considerable problems in their commercial buildings. Across the Indian Ocean, Australia's buildings and roads have been affected. China has indicated that 20 percent of its railroad system is built over expansive soil subgrades, and they too are doing something about mitigating the destructiveness.

Across the Pacific Ocean in the United States, buildings, highways, streets, utility lines, homes, sidewalks, and airports have all experienced the destructive impact of the swelling soils. The playa lakes in west Texas and in eastern Colorado have caused considerable distress to homes built on what was initially thought to be stable prairie soils. In other Rocky Mountain areas, front-range clays and shales in the Denver area have caused many class-action suits against builders. The Mission Viejo and Colonial Hills developments may have involved as many as 40,000 claimants, though one consulting engineer estimates that only 1 percent of the homes were seen as having suffered real damage. Similar class-action suits have been filed by home owners in Corpus Christi, Texas. A San Antonio, Texas, consulting engineer says the usual way of dealing with the swelling soils there has been to remove and replace the expansive clays. Lime treatment and pressure injection are also usual cures. There is a reluctance to use geomembranes because people are unfamiliar with their track record. In Arizona some home owners in the Phoenix area face expansive soil problems, but more home owners in the Tucson vicinity face them. In Tucson the dipping uplift strata are generally avoided, which is encouraging in view of the similar situations in two Colorado counties that have caused problems there.

Large commercial and industrial buildings are not immune to the challenges presented by the expansive soils. One of the first engineers in the United States to address the problem was Willard Simpson, Sr., in San Antonio, Texas. He first confronted the problem when his consulting engineering firm was dealing with the foundation design of a large office building that included the Smith Young Tower. Through correspondence with Karl Terzaghi and working with the Duderstadt Drilling Company, a new strategy evolved. A drilled-shaft foundation carried a reinforced-concrete shaft through the zone of moisture activity to a relatively stable material. A bell footing was sometimes used at the shaft founding elevation. On the Smith Young Tower, the drilled shafts provided the foundation, but Simpson was careful to set up a monitoring program and provide jacks to keep the building level.

Since then, the drilled-shaft foundation has become the usual foundation design mode for large industrial and commercial buildings where swelling clays are present. The drilled-shaft foundation has also become the frequent choice of bridge designers in the southwest. Yet some problems still arise with this foundation type. For example, on one large institutional building the shafts have actually been "grabbed" by the clay and torn apart. The solution has involved separating the shaft along its length from the clay. Thus the strength of the foundation will be secured from its founding depth.

## 5.2 Highways, Roads, and Streets

Highways, roads, and streets in the United States are among the major facilities impacted by the expansive soils. A study conducted by the Waterways Experiment Station (WES), U.S. Army Corps of Engineers (COE) for the Federal Highway Administration (FHWA) showed that 40 of the 50 states have expansive soils. Reports from state highway departments across the country have been published dealing with the challenge to their highways, roads, and streets. Fourteen states have reported using geomembranes in their efforts to control expansive soils on their highway systems. The U.S. Bureau of Indian Affairs has used the geomembranes on several reservations in the southwestern states. The COE reports using geotextiles on several roads they are responsible for in the southeastern states.

Colorado has been addressing the challenge of the expansive soils, frequently a Mancos shale, to their highway system. They have tried many different techniques. They probably were the first state highway department to use a Du Pont Company Typar geomembrane. They concluded that blasting and recompacting worked better for them. Colorado also has a continuing challenge in dealing with expansive soils in residential situations. There also has been some concern voiced about the impact of these soils to their new Denver International Airport. Reputedly, one of the consulting engineers recommended the use of geomembranes for a runway. However, the firm reviewing all the runway designs apparently decided instead to use the remove-and-replace option with an enhanced concrete pavement design.

The South Dakota Department of Highways has dealt with the expansive soils destructive movements by placing geomembranes vertically in an effort to arrest the moisture changes in the roadway subgrade. Of special note is E. B. McDonald's work [3]. The vertical barrier was placed 4 feet deep. McDonald stated that, had it been placed twice as deep, it probably would have stopped the pavement heaves and swells that highways and streets have suffered.

In Texas the cities of San Antonio, Corpus Christi, Waco, Dallas, and Fort Worth have all experienced the destructiveness of these soils, and they have reported on their remedial work. In Nevada the city of North Las Vegas has reported similar efforts.

## 5.3 Airfields

Airfields in the United States, South America, and Israel have reported on the challenge of the expansive soils. The Dallas–Fort Worth Intercontinental Airport has had to deal extensively with expansive soils. The new Denver International Airport and the San Antonio, Texas, International Airport have had to address the same expansive soils challenge. The study for the Federal Aviation Agency conducted by McKeen examines the soil conditions at six U.S. airfields, all with expansive soil subgrades. As with highways, the indications reinforce the opinion the swelling soils are widespread in the United States.

Suppliers have reported the use of geomembranes to control expansive soils at two airfields in Colombia, South America. This indicates the presence of these soils and their perceived impacts on airfields in north-central South America. More information is being sought. Livneh's paper "Israel's Experience With Runway Pavements on Expansive Clays" identifies four airfields in Israel that are built over expansive soils.

## 5.4 Railroads, Canals, and Pipelines

Railroads in China and canals and pipelines in the United States and Israel are also among the impacted. Papers presented by engineers indicate that 20 to 25 percent of China's rail system is built over expansive soils. They have reported on their successful efforts being made to control these destructive and expensive movements. Similar challenges have arisen in numerous rail lines in the United States, South Africa, France, and Russia.

Canals in California, including the Friant Kern, which is part of the central valley system, as well as those in several states west of the Mississippi River, have been impacted by swelling soils. Canals in Nebraska and Missouri have also been affected. In some important dike stabilization improvements, geomembranes have been used with success. These facilities are controlled by the Bureau of Reclamation, the first agency in the United States to use geomembranes [4]. The problem of leakage and seepage in unlined canals has also been faced in Oregon. In several test sections, geomembranes were used to remedy the problem [5].

Underground pipelines in both the United States and Israel have been affected by expansive soils. Information on both these canal systems is being sought. Home service lines—water and wastewater in particular—have contributed to the destructive action of expansive soils. In Texas, in the cities of Corpus Christi and San Antonio, breakage in these home service lines has been thought to have led to the destructive movement of the expansive clay foundation in the homes they serve.

## 5.5 References

1. Jones and Holtz. "Expansive Soils—The Hidden Disaster." *Journal of Civil Engineering Proceedings*, ASCE, August 1973.
2. Krohn and Slosson. *Assessment of Expansive Soils in the United States.* Fourth International Conference on Expansive Soils. Denver, 1980.
3. E. B. McDonald. *Experimental Moisture Barrier and Waterproof Surface.* South Dakota Department of Highways, St. Pierre, 1973.
4. U.S. Bureau of Reclamation. *Use of Geomembranes in Bureau of Reclamation Canals, Reservoirs and Dam Rehabilitation.* U.S. Department of the Interior, Bureau of Reclamation, Denver, December 1995.
5. Swihart, Comer, and Hayes. *Deschutes Canal Lining Demonstration Project.* U.S. Department of the Interior, Bureau of Reclamation, Denver, 1994.

# 6
# Testing

Testing is an important step in any geotechnical study, design, or construction. This chapter discusses test patterns, the number of tests to run, and the types of tests to conduct.

## 6.1 The Importance of Adequate Testing Prior to Design

The Denver newspaper April 1996 series of stories concerning a class-action suit involving almost 1000 home owners in the Mission Viejo subdivision is a contemporary reminder of the importance of testings. In that case there was a thorough investigation of the soil conditions prior to design. However, during the trial proceedings, it became apparent that considerable variation existed in the interpretation of the test results.

The evolution of the understanding of the expansive phase of soil mechanics began in the early 1900s. Atterberg developed a series of soil tests for determining shrinkage, plastic, and liquid limits. These tests are used to determine the plasticity index (PI), in which values over 25 indicate the possibility of an expansive soil. Dr. Karl Terzaghi's studies and publications beginning in 1913 established soil mechanics as a significant element of civil engineering. His decades of work rightfully earned him the title of the Father of Soil Mechanics. A review of early work in soil mechanics was presented at a symposium in Bangkok, Thailand, in 1994: *Developments in Geotechnical Engineering,* subtitled *From Harvard to New Delhi, 1936–1994,* by Professor Gdalyah Wiseman. Early work in the 1930s by Porter in Texas and Wooltorton in Burma were credited as they sought to understand the causes of damages to highways and buildings. In both locations the swelling expansive soils were determined to be the cause of the destructive results. These conclusions were reached regrettably in postconstruction testing.

Chen in his book reports over 1000 cracked buildings that he investigated [1]. He conducted his studies between 1960 and 1975 primarily on the front range of the Rocky Mountains in Colorado and Wyoming. Chen states that these continuing problems usually reflected a lack of prior design testing. He emphasizes the importance of adequate testing and recommends that the geotechnical engineer not accept a project where sufficient testing is not funded. A proliferation of class-action suits further underlines the wisdom of securing substantial testing prior to design. These legal actions are not limited to a particular section of the United States. They extend from the Texas Gulf coast through the midsection of the country and on to the Pacific Coast states. The financial aspects are important. The professional responsibility reaches far beyond the dollars involved.

## 6.2 Resources for Information on Soil Testing

Internationally the availability of resource material to alert the engineer to potential problems from expansive soils keeps increasing. From the 1930s conferences have led to the sharing of information on a worldwide basis. The early meetings were held at Harvard and later at Texas A&M University. They were followed by conferences at the Technion in Israel, Australia, and India. Supplementing these international conferences on expansive soils and unsaturated soils, regional meetings such as the Asian meeting held in Bangkok, Thailand, have provided additional reference sources. The Transportation Research Board of the U.S. Academy of Science offers more information on these soils in countries around the world.

Increased attention to and awareness of expansive soils have brought additional publications of books addressing the challenges. Though these publications sometimes focus on theoretical and laboratory testing, they provide further insights into the identification of the problem and potential solution to it. With the publication of papers presented at the meetings, substantial resource material is now available. The engineer and other members of the design and construction team should not ignore them.

In the United States the variety of sources to assist in alerting the engineer to the challenge of expansive soils is considerable. The expansive soil study by Waterways Experiment Station (WES) of the U.S. Army Corps of Engineers (COE) is a significant source. Headed by Donald Snethen and the late Lawrence Johnson, the project was funded by the Federal Highway Administration. It includes a summary of what states have identified the presence of expansive soils and reveals that these soils exist in over 40 of the 50 states. The WES study provides much more valuable information concerning theoretical and applied control efforts to manage expansive soils.

The Soil Conservation Service (SCS) of the United States Department of Agriculture has considerable information of soil conditions on a countywide basis. The building Research Advisory Board (BRAB) of the Federal Housing Administration provides additional nationwide assessments of the soils challenges. State agencies add to the resources available. The Colorado Geological Survey and the Texas Geological Survey are among those providing detailed information alerting engineers to the presence of the expansive soils. The U.S. Bureau of Reclamation and a wide variety of state departments of transportation have issued reports addressing the soil conditions in particular parts of their jurisdiction. City building codes are frequently another source of valuable information in dealing with these challenges. With all this material within reach, one might think that individual testing for a particular project could be considered superfluous and uneconomical. Little could be further from the truth.

It is worthwhile to reconsider Chen's advice that the geotechnical engineer not accept a project for which the client does not provide adequate funding for required soil testing. Colorado has a broad scope of soil testing reports and recommendations available. Yet with all those resources available, still in 1996 there were nearly 1000 class-action suits resulting from expansive soil damage. From the early works of Terzaghi, Casagrande, Peck, Porter, Wooltorton, and many others, the engineer is reminded of the need to check. Their recommendations for testing are repeated by, among others, Kassiff, Livneh, Komornik, and Wiseman; Lambe and Whitman; Seed and Woodward; Chen; and Nelson.

The critical elements to remember are the changing characteristics of soils. What is present in one place may not be present significantly several feet away. Helpful as all the books, studies, and reports are, however they offer only general warning signs. In fact, sometimes there are no warnings in any of the printed material, and regrettably, the engineer must rely only on his or her own direct observation of the ground at a particular site. In reading the geotechnical publications of the American Society of Civil Engineers, and of their GeoTech Institute, as well as the works of the International Society of Soil Mechanics and Foundation

Engineers (ISSMFE) and other engineering associations, the unfolding picture is that, while much has been learned about expansive soils, much remains to be discovered. The best protection for your professional reputation and for your clients is to test as much as needed.

## 6.3 On-Site Soil Inspection

The geotechnical engineer needs to make an on-the-ground inspection of the proposed construction site. The structures in the vicinity of the site should also be closely examined. Are there cracks in the building? Is the roadway distorted? Are the curbs and sidewalks showing signs of distress? Does the stream bank show a change in slope? What does the vegetation look like? These direct observations can be as valuable as the information gathered from publications. The data determined from the on-site inspection should be recorded because it will prove valuable in preparing design recommendations. It will also be valuable in discussing proposals with the cient and in answering any questions following construction.

The knowledge of the type of construction planned will further enhance the value of the site inspection. Certainly, the impact of expansive soils will be different for a nuclear power plant, a service road, a freeway, or a single-family residence. All of those projects require careful design, yet the funding for testing and preventive measures will vary among them. The clients' awareness and perceptions of expansive soils will also differ in degrees of sophistication.

## 6.4 Determining the Number of Test Sites and the Methods and Evaluating the Test Results

Determining the number of test site locations, the test depths, and which tests to be conducted are important in preventing expansive soil damage. Building owners frequently need to keep construction costs down, to be economical. However, this goal must be considered against the cost of repairing damages later. Taking an insufficient number of tests or not drilling deeply enough, may lead to inordinate postconstruction costs.

Frequently the field drilling will be handled by a crew using a mobile auger rig. The drilling crew should know the type of construction that is planned for the site, and they should be aware of the importance of the reliability of the information they are gathering. Logs should be kept of their work, and information should be clearly recorded.

If one is dealing with a single-family residential site, at least one and preferably two test holes should be drilled. An auger drill should be used, and the hole should be taken down 20 feet. The soils should be sampled every 5 feet or at intermediate intervals if there is a change in soil types. Larger structures will require drilling more test holes. For basements that will be in an excavation below natural ground, the depth of the test holes should be set accordingly. There's no point in drilling a 20-foot-deep hole if the structure's finished depth will be 30 feet below natural ground. The depth of any structure should be considered when drilling, whether the project is a multistoried building or a freeway in a depressed section.

On large planned developments, the test holes should be placed close enough to avoid the possibility of missing any significant soil condition change. For major commercial developments in a challenging soil condition area, the holes should be drilled 50 feet apart. In such cases taking soil samples every foot or two would be useful. On highway projects samples taken every half mile would be useful for initial testing. At bridge sites the number would vary according to the length and type of structure planned; however, a minimum of one at each abutment would be a sound starting point.

Open-pit holes are recommended for significant structural projects. This testing goes a step beyond the disturbed samples provided by the drilling rig. It provides the opportunity to take undisturbed soil samples for moisture and density testing. The Shelby tube is an intermediate-level test between the drilling rig's sample and the open pit. The open-pit test would provide the most realistic indication of the soil conditions. Field density tests in an open pit generally result in a truer measure of existing soil conditions. Determining the soil's Atterberg limits—its liquid and plastic limits—will provide some of the earliest laboratory indications of the expansive capabilities of the soil. The plasticity index, along with a sieve analysis and consolidometer and suction testing, will provide a clear assessment of the material.

Soil suction can be determined by using the filter paper procedure or by thermocouple psychrometer and tensiometer testing. Standard penetrometer and cone penetration tests provide an initial indication of the expansive characteristics of the sample. The Dutch cone penetrometer also provides information leading to soil characterization. Dilatometer and pressure meters are valuable tools in making determinations of an expansive soil. Nelson reports on Ofer's *in situ swell pressure device*, though this equipment is primarily seen as a research device. The

Atterberg limit tests often provide the basic field information to substantiate the soil's expansive nature.

The Atterberg limits PI values occur in a variety of source books as guides along with other tests as measures of swell capabilities. Kassiff indicates an extremely high degree of swell if the PI is greater than 32, high between 25 and 32, medium between 12 and 24, and low below 21 [2]. Chen's 1975 publication lists PIs over 35 as very high swell potential, 29 to 35 as high, 10 to 35 as medium, and 0 to 15 as low. He too lists other guidelines [1]. Nelson and Miller [3] use Chen's values from his 1988 publication of *Foundations on Expansive Soils,* although they are the same guidelines as those in Chen's 1975 edition [1]. Kassiff's additional testing results include swell in odometer values, the shrinkage limits, and free swell percentage [2]. Chen introduces Altmeyer's guide consisting of a shrinkage limit, with the linear shrinkage classified to the degree of expansion [1]. Nelson and Miller use Ranen's classification including the PI and shrinkage index [3]. There are additional tests used for the classification of these and other soil types. Care is, of course, a requisite in all testing.

## 6.5 Approaching the Design and Construction Stages from the Test Phase

If documentation of the testing procedures has been maintained, the study of existing conditions and the test results should be at hand. This will include all the steps enumerated previously in this chapter. The results will offer the engineer information on the options available for a particular project. The different tests, however, might provide a variety of results. The challenge will be to determine the most appropriate model.

It is useful to remember some of the experiences of geotechnical engineers. Chen reports using a half dozen different formulas with test results to compute potential vertical rise. The values ranged from 2 to 12 inches. When the structural engineer in a particular situation indicated that the facilities budgeted amount couldn't stand design for even the 2-inch value, Chen reevaluated. This led to a 1-inch heave estimate. Several years later, Chen tested the heave of the soil at the completed site. It was a ¾ inch. This recounting is not meant to denigrate the importance of the testing. Nor does it mean that the lowest calculated heave can be reduced by half and still keep the project on safe ground. What it does say is that soil mechanics is a technical art and that it should be practiced with an awareness that testing does not always yield precise formulations.

The testing must be done. It is a professional imperative. Moreover, the owner of the project must clearly understand the importance of it. At the same time, the owner must clearly understand that the testing is only part of the technical art of geotechnical engineering. The unknown is acknowledged while all the tools at hand are used. An assessment is made, and conclusions are reached. Yet all of nature's mysteries are not unraveled.

## 6.6 Selecting a Geomembrane

When, after careful and considerate analysis, it is decided to use a geomembrane, appropriate test methods are required for choosing the best product. The early TexDOT projects used the Du Pont Company's Typar specifications. No similar geomembrane was on the market in the United States at that time, and its use was characterized as experimental. When the geotextile was used in additional projects, a generic specification was developed.

The Wyoming Department of Transportation (WYDOT) developed their specification with heavy reliance on the American Society of Testing Material (ASTM) standards. The WYDOT specification is titled *Special Provision for Impermeable Plastic Barrier*. The tests specified cover six primary material requirements. They are wide-width tensile strength (ASTM D 4595), elongation at failure (ASTM D 4595), puncture strength (ASTM D 4833), permeability coefficient (ASTM D 4491), trapezoid tear (ASTM D 4533), and thickness (ASTM D 1777). Other items in this specification include construction, measurement, and method of payment.

The TexDOT Material and Test division (D 9) has used two specifications to address the placement of geomembranes and the material tests required. Their Special Specification 5646, *Vertical Moisture Barrier*, refers the material requirement to Texas Test Methods. The barrier specification is contained in the Materials and Test Division Department Material Specification, D 9 6210, *Vertical Moisture Barrier*, and it addresses many of the standards in the Wyoming specification. It uses applicable ASTM test methods in addition to several Texas Test Methods.

The Mississippi Department of Transportation in 1989 used the TexDOT model as its specification. Arizona has used a combination of specification items. For the engineer, the designer, and the builder, being familiar with the owner's requirements are most applicable in this phase of the testing program.

Geomembranes are studied within the growing geotechnical fabrics field. The growth of this field has culminated in the establishment of the Industrial Fabrics Association, the *Geosynthetics Report*, the North American Geosynthetics Society (NAGS), and Geosynthetics Conferences. Their information is very helpful. But the safest course to follow is to test, test, test.

## 6.6 References

1. Fu Hua Chen. *Foundations on Expansive Soils*, 2d ed. Elsevier, Amsterdam, 1975, 1988.
2. A. Atterberg. "On the Investigation of Physical Properties of Soils and the Plasticity of Clays." Int. Mitt. Boenkunde, 1911.
3. D. Snethen et al. "Technical Guidelines for Expansive Soils in Highway Subgrades." U.S. Army Engineer Waterways Experiment Station, Vicksburg, MS, 1979.
4. H. C. Porter, "Roadway and Runway Soil Mechanics Data." Texas Highway Department, Austin, 1942.
5. Soil Handbook for Soil Survey. Soil Conservation Service, U.S. Department of Agriculture, Washington, D.C., 1966.
6. Building Research Advisory Board. "Criteria for Selection and Design of Residential Slabs on Ground." Federal Housing Administration, Washington, D.C., 1968.
7. Colorado Geological Service. "Potentially Swelling Soils & Rock in the Front Range Urban Corridor." Department of Natural Resources, Denver, CO, 1974.
8. J. D. Nelson and Debora J. Miller. *Expansive Soils*. Wiley, New York, 1992.
9. G. Kassiff, M. Livneh and G. Wiseman. *Pavements on Expansive Clays*. Jerusalem Academic Press, 1969.
10. B. Seed, J. R. Woodward, and R. Lundaren. "The Mineralogical Aspects of the Atterberg Limits." *Journal of Soil Mechanics & Foundations ASCE*, vol. 90, New York, 1964.
11. K. Terzaghi and R. B. Peck. *Soil Mechanics in Engineering Practice*, 2d ed., Wiley, New York, 1967.
12. D. Wooltorton. "A Preliminary Investigation into the Subject of Foundations in the Black Cotton and `Kyatti' Soils of the Mandalay District, Burma." *Proceedings of the International Conference of Soil Mechanics and Foundation Engineering*, Harvard University, Cambridge, MA, 1936 (reprinted 1964).

# 7
# Design

The hazard potential must be addressed in the design stage of a construction project. There is a wide group of preventive strategies to choose from, enabling the engineer, with other members of the design team, to decide on the most effective and economical method. The decision is critical. Geomembranes in many cases, though not all, will provide a rational economic solution, and they are an important element to consider in design considerations.

## 7.1 General Design Considerations in Controlling Expansive Soils

Expansive soils pose different problems for residential, commercial, and institutional buildings and transportation facilities. However, the destructiveness of these soils among these different settings arises from the same cause: Moisture changes lead to volumetric changes in the soils. Geomembranes have been used in an increasing number of projects to minimize or prevent the damages caused by the expansion of the soils. The designer and the construction engineer have an increasing record of geomembrane projects upon which to draw, particularly in transportation facilities such as highways.

Most residences are fairly simple, one-story-high structures built on a variation of a pier foundation. This arrangement usually works quite well as it adjusts to the moisture change in the clays, and their expansion has little or no impact on the structure. Many multistoried residential units have also used the same foundation design with success. However, as construction techniques have evolved and the use of shiplap has been gradually supplanted by gypsum board, expansive soil foundation movements have begun to create more owner distress. Willard Simpson, Sr., in an article printed in the *Civil Engineering* magazine in 1934, states some basic facts to keep in mind when dealing with expansive soils [1]. For example, a soil may have one characteristic profile when undisturbed, but the same soil may be entirely different after it has been exposed to human activity. In dealing with the expansive soils in the San Antonio area, he observed that foundation areas turned from a previously firm clay to watery mud. He noted that some of the change was caused by the construction crew—working on the foundation and tramping through the mud. He designed ways to minimize this change and impact. For example, he found that using underdrains kept the buildup of water that was resulting from capillary action or, probably more frequently, from breakage of waterlines due to the soils' expansion. The underdrains fed into a sump pump or other positive outlet.

Two years later in 1936 Simpson spoke at the ISSMFE Harvard meeting of his 20 years of experiences with these expansive soils. Simpson and Duderstadt's development of the drilled-shaft foundation provided some solution to the destructive movements. The drilled shaft also worked for commercial and institutional buildings as well as a variety of transportation infrastructures, particularly bridges. The Building Research Advisory Board (BRAB), after considerable study and possibly stimulated by rising claims on federally insured mortgages, issued a report providing guidelines for residential construction. Their 1971 publication *Criteria for Selection and Design of Residential Slab on Grade* for the Federal Housing Administration took into account the diversity of foundation soil types and provided the standards for several decades. Builders and others came to conclude the BRAB design recommendations were unnecessarily demanding. The result was a study conducted at Texas A&M University for the Post Tensioning Institute (PTI) by Kent Wray, then a doctoral student, and Professor R. L. Lytton. Their recommendations became widely used as the PTI standards.

Another design procedure was developed using some of Chester McDowell's *potential vertical rise* (PVR) *methods*. The *cantilever beam design* (CBD) *approach* described in a paper by William Berger and John Phipps reported that between 1988 and 1993 homes built using this CBD in central Texas in expansive soil areas experienced no cracking. The CBD that evolved from the BRAB guidelines using the PVR involving Atterberg limits and load requirements apparently is achieving good results.

A historical question remains concerning Wooltorton's work in the Mandalay District of Burma. Taking place in the 1920s and 1930s, over 100 buildings were found to be suffering from considerable structural distress. Wooltorton determined that the cause of the buildings' problems were not construction shortcomings or design failures. Rather, he determined that the distress was caused by expansive soil. In his presentation at the Harvard meeting, he emphasized that proper drainage away from the buildings is crucial [2]. The need to keep vegetation with roots demanding water in dry time, finding and drawing it from underneath a building, is another effective measure.

Commercial buildings followed a similar pattern for single-story structures. The multistoried buildings were designed with the work of structural and geotechnical engineers providing vital technical data. As the weight of the structure increased with the multiple addition of the stories, there was more load to resist the forces of expansive soils. Simpson's early work provided the drilled-shaft foundation to support these multistoried structures. These foundations were used in the so-called skyscraper structures such as the Smith Young Building (later called the Transit Tower), and Nix Medical Office and Hospital Building in San Antonio, Texas.

The drilled-shaft foundation solution to the expansive soil problem, with footing depths to bedrock or past the active moisture zone, has become widely used. However, caution must be observed, and protective measures must still be taken in some cases. In one instance, an institutional building used a drilled-shaft design, but it still began to suffer significant structural distress. Investigation revealed that expansive soil had closed around the shafts and had succeeded in pulling the shaft in two. In certain cases it has become important to positively separate the sides of the shafts from the clay as it seeks a foundation depth beneath the level of the active moisture zone.

Institutional buildings have not been without their problems due to expansive soils. To deal with them, engineers have tended to design them according to the bearing capacity of the soils. This practice has prevented most of the more common problems. On many occasions, as noted particularly by Chen [7], designers rely on existing formulas for determining the load-bearing capacities of these pier shaft foundations. However, the uplifting forces and the consequences of using bell footing require careful further study. Frank Bryant, an Austin, Texas, consulting engineer, found that bell footing may actually cause more problems than it is solving. Pier uplift, skin friction impacts, and improper construction are also contributing factors in problem situations. In one particular case noted by Chen, a drilled pier was designed for a 20-foot depth, but actual postconstruction inspection revealed that it had been drilled only 4 feet deep.

Nelson and Miller review a variety of structural foundation options [3]. Included are drilled pier and beam foundations and stiffened slabs on grade, methods that were based on the BRAB, the Lytton, Walsh, Swinburne, and Post Tensioning Institute's procedures. The BRAB has examined four types of slab design: unreinforced, lightly reinforced against shrinkage and temperature cracking, reinforced and stiffened, and structural not supported on the ground. The reinforced and stiffened type is recommended for expansive soils. The principal assumptions are that the superstructure load is uniformly distributed over the slab, that soil pressure is uniform under the supported slab area, the value of the support index is independent of the slab dimensions, and the analysis of deflection and stress distribution is done as two one-dimensional cases rather than as a single two-dimensional case.

## 7.2 Expansive Soils Control Options

A variety of design methods exist to control expansive soils. The Lytton procedure was developed and reported in 1970, 1972, and 1973. It is summarized as calculating the maximum moment for each direction of a rectangle with the soil and footing rigid, the soil providing only line support. Nelson and Miller indicate that only three types of loads are calculated in this procedure. The first is the line load acting on the slab edge, the second is the line load at the slab's center, and the third is a uniformly distributed load from the interior live and dead loads. The maximum dimension of the structure using this method is determined to be 85 feet.

The Walsh procedure (1978) is based on a beam analysis on an elastic-coupled Winkler foundation. Using a parametric study of soil and structural variables, Walsh viewed the beam on a mound with three nondimensional parameters with support indices. Reportedly, Walsh concluded that the slab's shear strength was not an important design consideration. To implement the procedure, the engineer selects soil support indices from tables with design moment and develops stiffness calculations from the equations provided.

The Swinburne procedure was developed by two Australians, Fraser and Wardle, who used a finite element model of a slab as a plate resting on a semi-infinite elastic soil. The method relies on design charts that are provided to calculate moment, deflection, and beam depth for an assumed maximum differential heave, edge distance, concrete strength, and number and width of cross beams.

The Post Tensioning Institute procedure was developed by Kent Wray and Professor R. L. Lytton at Texas A&M University in 1978. Widely used and recognized in the Uniform Building Code, it is modeled as a soil structure interaction, a plate resting on an elastic continuum. Various sets of input variables are determined using finite element techniques; then a regression analysis is used to establish equations to calculate heaves at the center and edge of the slab. The calculations include design moment, shear, and deflection equations.

At an April 1994 seminar, *Experiences with Swelling and Collapsing Soils*, sponsored by Monash University, Melbourne, Australia, Professor Gdalyah Wiseman, now deceased, then of Israel's Technion, presented another design method, the *expert system* [4]. The system provides guidelines and design recommendations for use by a geotechnical engineer performing site investigations and proposed foundations. A *KnowledgePro* DOS is used to develop this expert system program.

The expert system is composed of four sections that, together, deal with site classification, laboratory testing, predicting and measuring swelling behavior, predicting heave, design considerations, and codes. A bibliography, photographs, and explanations are included as are answers to questions regarding laboratory testing and choices available for design. The system also describes how the recommendations are reached. The hypertext availability in the system offers photographs and explanations that are useful both to the engineer and to clients, as well as to other interested parties.

The site investigation phase is the first step in determining heave potential, needs for further laboratory testing, and directions for project development. The building environment section examines the sources of water, the types of heat generators that might be present to dry out the soil, and the waste containment structures on the site. An examination of existing structures in the vicinity often provides clues as to what might be expected from the foundation soils. All of these are factors that should be explained to the clients so that they can understand the risks.

The system's first program examines plasticity indices at 20-inch intervals and free swell at 5-foot spacing to a depth of 30 feet. The second program adds hydrometer tests, shrinkage limit determinations, and in situ density. The third includes further additions for determining swelling pressure in undisturbed soil samples and overburden pressure of percent swell. The fourth concludes with tests of undisturbed soil samples taken at 4.5-feet intervals to ascertain the soil's swelling potential. The soil types are evaluated by the results from the escalating level of testing directed by the expert system. Beyond the systems methods and recommendations, the authors restate the need to keep water used for vegetation away from buildings when swelling soils are present.

Regardless of the procedures used in the structural design, several items need careful consideration when the building is in an expansive soil area. Drainage has to be diverted away from the structure. There should be no ponding of water adjacent to the building. Vegetation also requires careful attention. Trees should be at least their full growth drip distance away from the building. Flowers and shrubs requiring watering need to be kept well away from the structure also. A review of *Home Construction on Shrinking and Swelling Soils* by Holtz and Hart for the American Society of Civil Engineers shows how a geomembrane can be used to keep the moisture changes away from the foundation of the new residence. Remedial work by Kirby Meyer, a consulting geotechnical engineer in Austin, Texas, made use of a geomembrane 8 feet deep around a residence. Professor Kent Wray, then at Texas Tech University in Lubbock, Texas, used a geomembrane similarly around two-story apartment units in Amarillo, Texas.

The transportation facilities most commonly facing the expansive soil challenge have been highways, roads, and streets. Porter's early work in the 1930s examined ponding on the highway between San Antonio and Seguin, Texas. This work was replicated on the west side of San Antonio in the 1960s. Dikes were built in a cut section and filled with water for 30 days. Following draining, the subgrade received a lime treatment, base, and an asphaltic concrete pavement. Monitoring the elevations and moisture levels continued during and after the process, and a zone of activity was determined as a result of the testing. Identification of the zone proved useful in later work involving deep vertical fabric moisture barriers.

The Mississippi Department of Transportation (MISSDOT) in the same decade also used a ponding treatment on an interstate highway in the Jackson, Mississippi, area. Under the direction of their research engineer Paul Teng, they drilled holes 20 feet deep at 5-foot centers. The holes were filled with a pervious material, dikes were built, and the area was ponded for 90 days. Twenty years later it was reported to be the only section of the interstate highway in the Jackson area that had not shown severe pavement distortion caused by expansive soils.

Conard Kelly, in the 1940s, initiated the use of lime on highways, streets, and roads in the United States. The percent of lime added increased as the plasticity index increased. The early pattern involved testing the subgrade to determine the amount of lime to be added, then scarifying the subgrade and spreading the lime from a tank truck. The lime was worked into the subgrade with a maintainer. Later a vehicle was designed and used that mixed the lime on the job with the subgrade. This treatment became widely used in Texas, Oklahoma, and New Mexico. As distortions continued on many roadways after lime treatment, the depth of the lime treatment was increased. The Oklahoma Department of Transportation (ODOT) increased the depth up to 2 feet. Some recent studies have indicated that high sulfate contents in the subgrade may result in the lime treatment's causing, rather than ameliorating, heave.

In Professor Dallas Little's 1995 *Stabilization of Pavement Subgrades and Base Courses With Lime,* a detailed discussion of how lime acts to stabilize pavement layers is presented in depth. The section on mixture design reviews three design procedures: the Thompson, the Eades and Grim, and the Texas. Thickness design and construction procedures are thoroughly examined as is life-cycled costing. Little credits Chester McDowell of the Texas Highway Department as the Father of Lime Stabilization. He also notes the work of E. B. McDonald with the South Dakota Department of Highways in reporting the long-term-strength effectiveness of the lime-stabilized subgrade in that state.

Little's book also reports on the use of lime slurries and, further, on a prevalent belief that the *lime slurry pressure injection* (LSPI) should be done only when the subgrade is dry or desiccated. Little cites references to disprove this thought. A Mississippi case, for example, was successfully injected with lime under wet Yazoo clay conditions [5]. And in the other corner, as it were, Nelson and Miller in *Expansive Soils* [3] present three references claiming that LSPI won't work unless the soil is dry [6].

*Lime Treatment At Depth,* a paper by C. M. Higgins, Research Report No. 41, Louisiana Department of Highways, June 1969, reports on three different methods to secure the mixing of lime with the subgrade. One was electroosmosis, the second a lime slurry high-pressure injection, and the third, the placement of lime in previously drilled holes. Neither the electroosmosis nor the placement of the lime in the previously drilled holes showed any signs of success. The pressure injection treatment showed the best results of the three. But the lime in the pressure injection method did not disperse throughout the subgrade. Rather, it was found to have penetrated generally between $\frac{1}{2}$ to $1\frac{1}{2}$ inches, from the injector tip. At the end of 4 years, no lime was determined to be available for further treatment of the subgrade.

Lime has been used as a stabilizing agent in foundation soils for centuries in China and other places around the world [7]. Chen observes that lime lost support as a stabilizing agent in favor of cement, which showed a greater increase in strength earlier than the lime. Since strength is not as important as reducing the expansive soils' tendency to swell, lime should be the material of choice between the two additives according to Chen. Many engineers agree with Chen, as evidenced in the use of lime by state highway departments and in the Dallas–Fort Worth International Airport, which used 300,000 tons of lime in its construction.

In addition to the lime stabilization methods, other chemicals and materials have been used in efforts to control expansive soils. They include cement stabilization, asphalt stabilization, combinations with fly ash such as lime and fly ash and cement and fly ash, and a variety of patented chemical combinations. Electroosmotic chemical stabilization has been tried by the Arizona DOT with the use of potassium to make the expansive clay less active. Avoiding the expansive clay areas is a good, if not largely impractical, idea. Heavy asphalt applications extending to encapsulization, such as that used by W. J. Van London of the Texas Highway Department on a Houston freeway, has been tried. B. A. Brakey's work for the Colorado Highway Department with asphalt applications of 1 gallon per square yard led to the first use of a geomembrane on a state highway project.

Considerable attention has been focused on the design of geomembranes in expansive soils areas. The evolution of the use of geomembranes in efforts to control the destructive movements of expansive soils provides the guidelines for design and construction decisions. The phrase "if it works, don't fix it" remains applicable here. The early projects on Texas highway contracts used the Du Pont Typar material. As there was no known competitive geotextile being marketed at the time, the geomembrane was considered experimental. The Du Pont specification was used both on the General McMullen Drive rehabilitation contract and the IH Loop 410 work on the west side of San Antonio, Texas. It is interesting to consider the performance of the lightweight thin geomembrane on its horizontal placement on the General McMullen Drive. Despite trucks' traveling on the fabric, dumping base on it, and the base being worked with maintainer, rollers, and water trucks, no tearing took place.

On the Texas contract, where the geomembrane was first placed vertically, the geotextile exhibited none of the material negative aspects. The company's "experimental" specification again proved satisfactory. Now several companies are involved in supplying geomembranes. It is a competitive bid environment, and state agencies have developed their own or American Society of Testing Materials (ASTM) specifications, or they use a blend of both. It would seem that these specifications would be applicable for transportation facilities and buildings. It is beneficial to all concerned to keep current with the literature. Several state specifications are included in the appendixes of this book.

## 7.3 Geomembrane Placement Design Considerations

The placement of the geomembranes—horizontally, vertically, and horizontally and vertically—are design and construction concerns. In the case of a highway transportation facility, it is noted that the first placement of a geomembrane in Colorado was horizontal. It followed the pattern of a heavy asphalt treatment of the subgrade. The first Texas Highway Department geomembrane project was on General McMullen Drive, an arterial city street. Two other THD city projects placed the geomembrane horizontally.

One powerful reason for horizontal placement is the fact that the urban utility lines are seldom located as shown on the plans. Thus, extreme caution is needed in digging vertically a set distance from the pavement. On the other hand, there is a growing awareness that pavements can no longer be considered substantial water barriers. Studies by Barry Dempsey and Ernest Barenberg at the University of Illinois have supported this finding. Thus, there is good reason to use horizontal placement for the geomembrane. When it comes to dealing with geomembranes and buildings, part, if not all, of the placement extending from the structure must be horizontal.

Vertical placement of a geomembrane has been widely used in the United States and Australia. The objective is to provide a barrier from the surface to the depth of the moisture zone of activity. In Texas the placement at an 8-foot depth evolved from moisture studies prior to and after a ponding project on U.S. 90 on San Antonio, Texas's west side. There it appeared that major moisture changes took place from the surface down 6 to 8 feet. This tended to substantiate E. B. McDonald's observation from his early vertical barrier test on a South Dakota highway. In that situation, the fabric was placed 4 feet deep along the roadway shoulder with a pronounced reduction in heaves compared to an adjacent section of pavement without a geomembrane. McDonald theorized that if the geomembrane had been placed twice as deep, it would have stopped the heaving and swelling of the pavement. On highway projects there are substantial reductions in maintenance costs where a geomembrane has been placed vertically. In Texas and other states, this position is 8 feet deep. Wyoming reports success with its placement at only half that depth. VICRoads in Australia has indicated that their active zone of moisture is in the range of 2 to 2.5 m ($6\frac{1}{2}$ to 7 feet). The determination of how deep to place the fabric should be based on an assessment of the zone of activity.

Placing geomembranes both horizontally and vertically is done in both residential buildings and highway facilities. In the case of residential structures, the pattern has been recommended in a booklet distributed to Colorado home owners. The booklet recommends that the clay be prevented from opening up cracking patterns during dry periods to avoid becoming waterways during wet periods. This infusion of moisture leads to the volumetric changes that cause heaves and swelling action in these clays. This pattern of placement has been used in Texas in the remediation work by Kirby Meyer in the Austin area and on the large Amarillo, Texas, apartment rehabilitation work by Kent Wray.

In the highway field, the pattern of placing geomembranes both horizontally and vertically is being used on more than 50 Wyoming projects. The Wyoming Highway Department has considered the results of this method to be favorable and has continued to use it, accounting for more than 4 million square yards of geomembranes installed since the early 1990s. The combined horizontal and vertical placement reduces the impact of water intrusion through the pavement as well as from the side. Placement is the designer's decision. Using a horizontal and vertical pattern does add a significant amount to the estimate. A reduced unit cost, though, is a benefit of such placement, even though the total cost is higher with the probability of life-cycle costs being less.

## 7.4 Applicability and Feasibility of Various Design Approaches

In the realm of residential units, the recommended placement pattern is horizontal, moving away from the structure, then vertical. This seems to be a rational accepted pattern worth following. For highway facilities, the question of whether to use the geomembrane horizontally or vertically or a combination of both in a single placement are design decisions. First-cost and life-cycle expenditures are factors to weigh in the decision-making process. Other questions may be raised as to how much the existing pavement is retarding the entrance of precipitation during rainfall periods. Whatever the design decision, records on existing projects are available to be consulted as needed. One fact to stir into the decision is that the cost of geomembranes has decreased over the years while installation production rates have gone up. The issues of whether to use an underdrain in the bottom of the geomembrane trench and what backfill material to use require careful consideration. The thickness and other specifications of the geomembrane are established largely by a variety of state and ASTM standards. To a significant extent, design can be guided by previously completed projects.

A recent Australian study has cast additional light on the question of where to place the moisture barriers [8]. This study indicates that the life of a highway is usually considered to be 20 years. Roadway failure is generally attributed to structural loss of the pavement by rutting, cracking, or potholing. With an expansive soil subgrade, however, a life of 5 years or less is usual. The question then becomes when and where to rehabilitate the highway. Basically, problems arise from the intrusion of water into the subgrade, resulting in volumetric changes, heaves, and swelling in the pavements. The measurable results are wavelength roughness.

In the Australian study [8], expansive soils have been identified in destructive subgrades in half the area of their state, Victoria. The Thornthwaite moisture index (TMI), with ratings of +5 to −20, is a climate rating system that was used in the identification of expansive soils areas in Victoria. In their research, Mann and McManus identified *gilgai* as the major cause of long-wave roughness in their pavements. The researchers state that these gilgai, bowl-shaped depressions in the subgrade, return even after they are bladed level in roadway construction.

Roughness in Victoria has frequently been identified by visual observations and centerline profile indices. However, a more definitive method was sought, which led to the development of the *power spectral density* (PSD). This test uses many sinusoidal curves, wavelengths, and amplitudes phases. A Fourier transform is used to compute the amplitudes of the sinusoids, resulting in a profile when they are added together. When the transform is adjusted with the sinusoids, the PSD is developed. As the wave number (that is, the inverse of wavelength) increases, the PSD decreases. The mathematicians may follow this further. The result is that there is a quantitative method with which to compare a smooth riding section of highway with a suspected rough section.

The *international roughness index* (IRI) was designed as a general measure of pavement condition, although the PSD is still a more definitive value in assessing the need for a barrier and, if needed, its position. Research from the Queensland University of Technology has indicated that the wavelength most affecting truck ride and the frequency of heaves is in the range of 10 to 20 m (32 to 65 feet) depending on whether the truck has steel or air-bag suspension. The comparison of the PSD of the smooth riding section with the rougher segment gives a visual indication of where the barrier needs to be placed and the pavement resurfaced.

A report in the 1165 series raises several interesting questions [9]. It includes a list of situations where vertical moisture barriers are not effective. The report states that vertical moisture barriers are not effective in extremely dry climates, in ponded drainage conditions, or in tightly cracked soils and that they should be placed at a depth at least equal to the root depth. It further concludes that geomembranes are effective as vertical moisture barriers with the limitations listed previously. University-based studies are most helpful in the design of moisture barriers. This study's conclusions, based on 6 of the 24 projects within the Texas highway system, certainly deserves careful consideration. Engineers in Australia reviewed these limitations, but they continued with their plans for a geomembrane vertical moisture barrier since they concluded that none of the limiting conditions existed. Another element to consider in the decision process is the cash savings noted in the many geomembrane projects in the San Antonio, Texas, area where some of the limitations did not exist.

No discussion of design factors to consider when dealing with geomembranes and expansive soils should fail to mention the key elements of vegetation and grading. When not using geomembranes or other significant moisture barriers, these elements become even more critical in minimizing the opportunity for moisture change in an expansive soil. However, even with the use of vegetation, a moisture barrier should be installed. Moreover, trees planted too close to a house can increase the risk of root penetration in a variety of foundation situations. A recent publication by the Colorado Geological Service mentions that a minimum of 15 feet be maintained [10]. For shrubs and flowers, they suggest that a minimum of 5 feet to the house walls be maintained. Grasses requiring substantial amounts of water are best avoided. Watering or irrigating in a situation that creates a perched water table can also pose substantial problems. Good positive drainage by grading away from the building should be created and maintained. Ponded areas should not be permitted near buildings. The Colorado publication lists those grasses, shrubs, flowers, and trees that are most compatible in a swelling soil situation. Mulch gardens and Xeriscaping (a water-conserving method of landscaping) are highly recommended. Treating service lines improperly can limit the ability of the moisture barrier to give the needed protection and therefore require extra care.

Geomembranes are generally effective as moisture barriers. Though vegetation considerations should be carefully followed, Australian studies indicate that the barriers can deflect the movement of the roots. To secure this outcome, they recommend a geomembrane vertical moisture barrier, properly placed, as an effective means of controlling expansive soils.

## 7.5 References

1. W. E. Simpson, Sr. "Foundation Experiences with Clay in Texas." *Civil Engineering*. ASCE, New York, November, 1934.

2. D. Wooltorton. "A Preliminary Investigation into the Subject of Foundations in the Black Cotton and `Kyatti' Soils of the Mandalay District, Burma." *Proceedings of the International Conference of Soil Mechanics and Foundation Engineering*. Harvard University, Cambridge, MA, 1936.

3. T. D. Nelson and D. H. Miller. *Expansive Soils, Problems and Practice of Foundation and Pavement Engineering*. Wiley, New York, 1992.

4. G. Wiseman, G. Zeitlen, and A. Komornik. *An Expert System for Foundations on Expansive Soils.* Seventh International Conference on Expansive Soils, Dallas, 1992.
5. *National Lime Association Bulletin 331,* Arlington, VA.
6. C. M. Higgins. "Lime Treatment at Depth Arlington, Va." Louisiana Department of Transportation. Research Report No. 41. Baton Rouge, 1969.
7. G. G. Ingles and R. C. Neil. "Lime Grout Penetration and Associated Moisture Movements in Soil." CSIRO Australia, Division of Applied Geomechanics, 1970.
8. L. Lundy and B. J. Greenfield. "Evaluation of Deep in Situ Soil Stabilization by High Pressure Lime Slurry in Section." Highway Research Record, Washington, D.C., 1968.
9. Fu Hua Chen. *Foundations on Expansive Soils.* Elsevier, Amsterdam, 1975.
10. R. Mann and K. McManus. "The Choice of Barrier Locations." *The Management of Moisture in Expansive Soils.* A seminar, Swinburne University of Technology, Melbourne, Australia, April 1997.
11. R. Javatilaka, D. A. Gay, B. L. Lytton, and W. K. Wray. *Effectiveness of Controlling Pavement Roughness Due to Expansive Clays with Vertical Moisture Barriers.* Research Report 1165-2F. Texas Transportation Institute, Texas A&M University.
12. D. C. Noe, C. L. Jochim, and W. D. Rogers. *A Guide to Swelling Soils for Colorado Homebuyers and Homeowners.* Colorado Geological Survey, Department of Natural Resources, Denver, 1997.
13. Building Research Advisory Board. "Criteria for Selection and Design of Residential Slab on Grade." Federal Housing Administration, Washington, D.C., 1968.
14. W. K. Wray and R. L. Lytton. "Design and Construction of Post-Tensioned Slabs on Grade." Post Tensioning Institute, Phoenix, 1980.
15. W. G. Holtz and S. S. Hart, "Home Construction on Shrinking and Swelling Soils." Colorado Geological Survey Special Publication 14, Denver, 1978.

# 8
# Case Studies: Buildings

A review of the areas where buildings are impacted by the expansive soils will be presented along with a variety of methods used to reduce the damages caused and the cures employed. In several cases geomembranes have been used to effect the cure. Other basic factors have been introduced presenting additional measures to be taken regardless of the conditions employed.

## 8.1 Expansive Soil Experiences in the United States, South Africa, Canada, Australia, and Israel

In the early days of this century geotechnical engineers in the United States, South Africa, Canada, Australia, and Israel began to study expansive soils. Willard Simpson, Sr., authored an article in the November 1934 issues of the American Society of Civil Engineer's *Civil Engineering* magazine entitled "Foundation Experiences with Clay in Texas" [1]. A structural engineer who headed a well-respected firm in San Antonio, Texas, Simpson recounted the problems and damages known to have been caused by these clays to residences, schools, hotels, and office buildings. Simpson was a 1905 Massachusetts Institute of Technology graduate, and he had professional interests extending into the geotechnical and foundation areas. His contacts and consultations with Karl Terzaghi spanned decades. Simpson's 1934 article asserted that a $\frac{1}{4}$- to $\frac{1}{2}$-inch movement in a one-story structure could be destructive. In a two-story building, however, it wouldn't be a problem. He further observed that a 4- to 5-inch movement in a large multistoried building would not be viewed as serious.

Two years later at a meeting of the International Conference on Soil Mechanics and Foundation Engineering (ICSMFE) at Harvard University, D. Wooltorton, executive engineer, Public Works Department, Meiktila, Burma, India, presented a paper addressing the impact of the expansive soils on buildings in Burma. (Actually, the publication by the ICSMFE is described as an "abstract" of the full paper.) Based on his investigation of over 100 building failures in the Mandalay District of Burma, Wooltorton states that the building foundation in such clay areas should extend down to the stable moisture zone. His study revealed that the structural design and construction work did not contribute to the distress shown in the buildings. Rather, the culprit was the expansive soils in the area. The area had experienced high rainfall periods along with prolonged dry spells, which resulted in the buildings' damage, with some cracks extending more than 10 feet deep. Wooltorton's abstract includes moisture readings in the district extending from 1923 to 1935. The readings are further tabulated to show rainfall data in the months of September, October, and November, which are often characterized as the "monsoon season." Additionally, chemical and mechanical analyses of over one dozen elements and compounds are presented in the paper.

Wooltorton includes a substantial number of photographs in the paper. They show the wide range of buildings affected by expansive soils in the Mandalay District prior to the middle 1930s. The buildings range from large government structures such as the post office, the civil hospital, the Mandalay Jail, the Fort, and the Agricultural college. Data tables are included as well as sketches of the buildings showing where the cracks have occurred. In his studies, Wooltorton observed that the movement of the expansive clays was both horizontal and vertical, and he emphasized that the damages to the buildings were due to these soils, not to design or construction shortcomings. He made several recommendations to prevent the types of damages sustained in Burma. They included the use of drainage systems and impervious aprons to minimize the seasonal moisture variations that cause cyclical and bearing changes. He reported on the existence of similar soil challenges in India, Sudan, and South Africa. The abstract and the complete version of the paper provide substantial and valuable geotechnical information [2].

Willard Simpson, Sr., attended the meeting and participated in the discussion following Wooltorton's presentation. Simpson supported Wooltorton's findings and cited his own 20 years of experiences during which he had found a similar need to place the foundation footings in a stable moisture zone. He further recommended that, on pier and beam foundations, the beam be formed on three sides. The beam should not be poured directly on the clay subgrade. He suggested that the shape of the bottom of the beam should preferably be triangular rather than horizontal and flat. This practice would reduce the uplift pressure of the expansive soil.

Another part of Simpson's legacy is the drilled-shaft foundation, which he developed with Ed Duderstadt, a water well contractor. This foundation method became widely used for buildings, particularly in the western United States. It is used to this day not only in buildings but also in bridge foundations. Simpson, with his firm's chief engineer, Manfred Gerhardt, and in consultation with Terzaghi, used a mat foundation in the clay foundation of the Gulf Building in Houston, Texas. This presented an economical switch from the pile foundations used in clays. The building was financed largely by Jesse Jones, later a U.S. Secretary of Commerce, who asked during construction for Simpson to add two more floors to the structure. Jones said he would hold the phone while Simpson checked out the possibility. After some rapid calculating, with Gerhardt, Simpson replied yes. In 1997 the ASCE recognized the Gulf Building as its National Historic Civil Engineering Award recipient.

Concern was increasing globally on the impact of expansive soils on buildings. The Federal Housing Administration, in cooperation with the National Science Foundation, published in 1971 the Building Research Advisory Board (BRAB) design criteria titled *Criteria for Selection and Design of Residential Slabs on Grade*. The intense work preceding this publication was a further indication of the mounting concern about the building damages caused by expansive soils. Fu Hua Chen, in his book published in 1975, reports on a study of over 1000 structures suffering from expansive soils damages [3]. Chen recounts these problems primarily from Colorado's Front Range area and Wyoming.

Chen's book was followed in October 1978 by the publication of *Home Construction in Shrinking and Swelling Soils*. The authors were Wesley Holtz and Stephen Hart, who wrote for the American Society of Civil Engineers, which had been awarded a grant by the National Science Foundation for the study. The publication was retitled "Information for Colorado Homeowners," (Special Publication 11) and was distributed by the Colorado Geological Society. This small booklet had much valuable information for the home owner, the engineer, the contractor, and builder. Its introduction notes that in 1978 damages caused by expansive soils in the United States cost $2.3 billion annually. In 1978 dollars, these damages amounted to twice the damages caused annually by tornadoes, earthquakes, flooding, and hurricanes together. Colorado's soil and climatic conditions combine to make it area prone to damages caused by expansive soils. The booklet explains what the soils are, where they occur, and what the owners' responses need to be to prevent damage from them. With photos and diagrams, it attempts to advise the home owner how to face the challenge. It is of interest to consider the drawings showing sheetings attached to exterior walls and extending horizontally from the home. Of equal interest is another drawing showing a plastic membrane extending horizontally away from the building, and then turning down vertically. Coauthor Holtz had served as chief of the soils and research sections of the U.S. Bureau of Reclamation, which was the first governmental agency in the states to use geomembranes.

A paper presented in 1985 at the Fifth International Conference on Expansive Soils examines the results of a test at Fort Collins, Colorado. The four buildings in the study had pier and grade beam foundations on an expansive Pierre shale formation. All four had a geomembrane placed horizontally under the floor slab that was separated from the subgrade that had 30 cm of lime treatment. Two of the buildings were in an irrigated environment, the other two, nonirrigated. The Pierre shale had an average liquid limit of 60 and average PI of 27, ranging up to 47. The geomembranes were placed 2.5 m (7 feet) deep, one on an irrigated site, the other nonirrigated. The report concludes that the pier and beam foundation performed well, but the lime treatment of the subgrade seemed to offer no benefits. The vertical geomembranes were not effective in reducing the total heave. They were, however, effective in reducing the total moisture migration and heave and the differential heave, as well as in producing a more stable moisture environment [4].

In 1997, Special Publication 43 of the Colorado Geological Survey provided more information to the Colorado homeowners [5]. The information is presented in greater detail than the publication of 20 years earlier. The 1997 guide covers the geology of swelling soils, subsurface moisture, construction on swelling soils, landscaping on swelling soils, home maintenance on swelling soils, swelling soils and homeowners' risk, and checking a property for swelling soils. This guide is bigger than the previous edition and presents timely information.

The booklet's summary of chapters (Part I) provides a good encapsulation of what follows in greater detail. Chapter 1 states that swelling soils and bedrock contain clay minerals that attract and absorb moisture. They swell when they absorb the moisture, and they shrink when they dry out. The chapter explains *bentonitic,* a term frequently used when talking about swelling soils. Some bentonitic layers are weathered volcanic ash which becomes very expansive clay. Measures are identified to control the swelling soil. Expansive soils are liable to be present on almost all Colorado properties except those at the highest elevations. The evaluation of how much swell can be expected depends on careful soil analysis of each property. Layers of steeply dipping mountain bedrock are identified particularly as posing special foundation challenges.

Dipping bedrock has attracted considerable recent attention in Colorado [6]. This publication indicates the heaving from these formations can be more severe than the damaging effects from the flat-land expansive soils. The *Dipping Bedrock Overlay District* (DBOD) was established in Douglas County, Colorado, and it is adjacent to the southern border of Denver's Arapahoe County. To Douglas County's western boundary is Jefferson County, where the dipping bedrock challenge also appears. These bedrocks are predominantly Pierre shales, but they also have Carlile and Graneros shales, a Fox Hill sandstone, a Greenhorn limestone, and Niobrara and Laramie formations. The DBOD was formed in response to the expanding building developments brought on by growing population demands. The counties involved and members of the investment and development groups are recognizing the financial implications of these geological challenges.

Despite these and other efforts, building damages continue to occur in the Denver and Front Range areas in Colorado. Feelings developed that the BRAB criteria were too conservative. In response, the Post Tensioning Institute (PTI) funded studies leading to the development of their less conservative design standards. Professors Kent Wray and Robert Lytton were major contributors to this analysis. The PTI standards became widely used in residential designs, but they did not become a panacea.

Two prominent geotechnical engineers in Texas used geomembranes in several residential projects. Kirby Meyer, president of MLAW in Austin, Texas, designed almost 50 projects using a geomembrane placed first horizontally away from the building, then vertically 8 feet deep. When the topography permits, an underdrain pipe is placed at the bottom of the geomembrane trench. The pipe discharges away from the building, which will reduce the foundation's subgrade moisture change and the clay's tendency to experience the volumetric change that results in cracking and heaving of the building. Wray, then in Lubbock at Texas Tech University, now dean of engineering at Ohio University, Athens, Ohio, has used geomembranes in a similar fashion on buildings. In the 1980s Wray used geomembranes as part of a rehabilitation project of a two-story, 18-building apartment complex in Amarillo, Texas. The complex was constructed on a site subject to flooding during heavy rains. As drought conditions prevailed for significant periods, the heavy rain that followed brought runoff to the many cracks in the clay. This resulted in swelling of the building foundation soils, the hanging of windows and doors, and the cracking of walls. Wray proposed the use of geomembranes horizontally and vertically. All the buildings behaved well following the geomembrane remediation except one.

A 1996 publication by Richard Handy of Iowa State University titled *The Day The House Fell* examines some of the challenges swelling and shrinking soils present the home owner. It indicates an awareness of expansive soils concerns through the midsection of the United States and chronicles the spate of class-action suits in California, Colorado, and Texas. All of the suits involve damages to buildings, usually residences, caused by expansive soils.

In South Africa, a distinguished worldwide leader emerged in the efforts to control expansive soils. Dr. Jeremiah E. B. Jennings, a professor at the University of Witwatersrand and head of the country's Building Research Council, addressed the challenges to the total range of infrastructure, including buildings. Expansive soil damages to residences and commercial buildings there are considerable. A. A. B. Williams, G. E. Bight, Van Der Muerve, and Z. Ofer are among the other notable engineers in South Africa who are addressing the need to reduce the damages caused by these soils. (Blight and colleagues' work with preheaving is reviewed in an earlier chapter.)

Canada's engineers have faced the similar challenges. More common in the western provinces east of the Rocky Mountain chain, there has been and continues to be efforts to diminish the expansive soils' destructiveness to buildings. Leaders in these efforts have been Professors Morgenstern, Leonards, Fredlund, and Bathhurst. From east to west across their country, the focus continues to be on understanding and ameliorating these destructive impacts.

Australia is contributing mightily to the ongoing global efforts to control expansive soil damages. Gordon D. Aitchison, chief of the Division of Applied Geomechanics, Commonwealth Scientific and Industrial Research Organization, Sydney, Australia, is another world-renowned leader in this discipline. Substantial damages have accrued to buildings in Australia, especially in the most heavily populated areas. F. W. Mitchell, J. E. Holland, and J. H. Holden are also providing leadership in Australia's efforts to protect their infrastructure. Working with Swinburne University of Technology, VICRoads and Holden continue to address the stabilization of the buildings damaged by the swelling soils.

After two conferences on expansive soils at Texas A&M University, the third meeting was held in Haifa, Israel, at that country's premier engineering university, the Israel Institute of Technology, the Technion. Buildings in Israel suffering damages from expansive soils are scattered throughout their small country. Israel began addressing the challenge of these soils in the late 1950s, led by J. G. Zeitlin who was joined by Amos Komornik, G. Wiseman, M. Livneh, S. Frydman, J. Uzan, and Eng. S. Nisichi. Building damages there remain substantial despite control efforts, though increased care is bearing encouraging positive results.

Many other countries face similar damages from these destructive movements, among them, India, Mexico, Colombia, Venezuela, Peru, Saudi Arabia, the Gulf States, Poland, and Russia. The report from the Arabian peninsula of severe cracking at one of the Saudi king's palaces caused by expansive soils confirms the universality of the challenge. They doubtlessly will be joined by others yet to report on their efforts to control the expansive soils.

## 8.2 New Construction and Remediation

New construction and remediation have seen a wide variety of methods used for dealing with the destructive forces of expansive soils in the United States. The drilled shaft is now being used on many residential, as well as commercial, buildings. The BRAB and PTI design guides have been employed in the construction of thousands of residential buildings. In the Austin and Amarillo, Texas, areas, Kirby Meyer, president of the MLAW consulting engineering firm, in the BRAB, and Professor Kent Wray in the PTI, have used geomembranes in remediation work (Fig. 8-1). Meyer's approach has been to extend the geomembrane first horizontally away from the building, then to run it 8 feet deep vertically around its perimeter. The Austin area has considerable amounts of swelling clays. Where possible, in the MLAW designs, an underdrain is placed at the trench bottom, and its outfall is placed well away from the structure. Meyer estimates that this method has been designed for use on almost 50 buildings. The geomembrane is a coated polyethylene, which provides a generally impervious barrier to water. Frequently the design includes an underdrain pipe. Where topography permits, the underdrain has a positive outlet.

Wray, now dean of engineering at Ohio University, has used a geomembrane similarly in the Texas panhandle area. The project involved an 18-unit apartment development that was constructed over an expansive soil subgrade. A prolonged drought period followed by a period of heavy rains resulted in area flooding. The buildings suffered considerable distress with heaving and swelling soils acting on the foundations. However, underpinning and mudjacking brought the two-story buildings into a plumb condition with windows and doors again moving with ease. Wray was asked to design a method to prevent the problem from happening

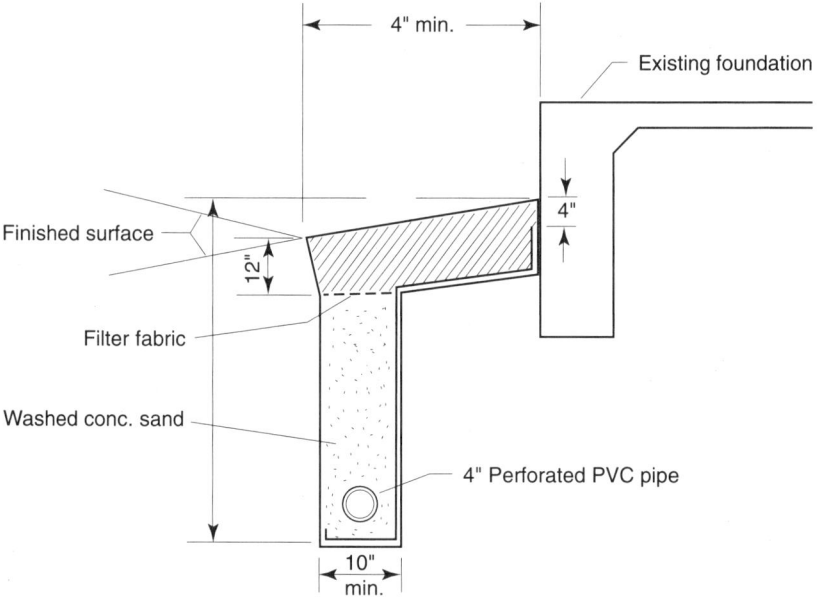

**Figure 8-1.** Kirby Meyer's Remediation Design—subdrain with vertical moisture barrier.

again. He chose a geomembrane placed 12 to 18 inches deep along the buildings' perimeter beams. The geomembrane was extended horizontally 3 to 5 feet away from the structure and then turned vertically 12 to 18 inches deep along the buildings' perimeter beams.

Over a 5-year period, all the buildings behaved well but one. In that particular building, the doors and windows were binding. For a while, the owner resisted doing anything further. Then one day an offer was made to purchase the apartment complex, and the owner decided to have Wray investigate further so that he could sell it. Excavation was done initially around the grade beam and the piers. The 12-inch piers were found to be in contact with the grade beam. Next the water service line was pressure tested. No leakage was determined. Checking the wastewater line was not feasible for pressure testing, because that line carries a low volume of water, it was not a significant contributor to the problem. Excavation was carried down to the bottom of the pier. There it was found the pier underreaming was not made as called for in the plans. The only other contributing factor relating to the building's distress was a large mesquite tree in the front yard. Vegetation can create quite a problem for foundations built on expansive soils, a discovery made in both the United States and Australia.

Buildings

Australia has undertaken a long-term remediation project involving six building sites in the state of Victoria. This work has been described in a paper by James C. Holden, in which he addresses not only the movement of the roots in their search for water but also their impact on building foundations in expansive soil areas [7]. The tests began in 1990 and have been conducted on developments in Victoria. The building foundations were dug in expansive clay where moisture changes result in volumetric changes, leading to heaving and swelling (Figs. 8-2 and 8-3).

**Figure 8-2.** Expansive soil and building wall.

**Figure 8-3.** Pavement cracking over expansive soil.

Holden reports on the previous efforts to create barriers with reinforced concrete and a variety of other materials. The report cautions against using chemical barriers as their life expectancy for blocking root growth isn't known nor is their impact on the environment known. Another disadvantage of chemical barriers is that they cannot block moisture changes. The report cautions against using other materials such as aluminum, corrugated iron, corrugated plastic, and asbestos. In his studies of these materials, Holden found nine "nots" to explain why they didn't work. They are not: high enough, deep enough, long enough, impervious to roots, strong enough, sealed against service lines, able to prevent roots from growing underneath, flexible enough, and cost effective.

Since 1990 VICRoads has been working to apply what they've learned about the use of geomembrane vertical moisture barriers to building problems such as that seen in Fig. 8-4. In 1990 VICRoads was asked to assist in the rehabilitation effort on a building on Hoddle Street in Collingwood, Victoria. It was decided to install a geomembrane 2 m (6.5 feet) deep, chosen as the maximum depth of moisture buildup and drying out in the Victoria area. This depth exceeds the cracks appearing during the drying periods that later provide passageways for roots and, during wet periods, for water. A modified chain trencher excavated a ditch 150 mm (6 inches) wide. When the clay became very wet, powdered lime was sprinkled on the chain teeth to keep the caked clay from

**Figure 8-4.** Expansive soils impact on Australia.

Buildings

coating the teeth and slowing down the operation. A backhoe was used to excavate around the service lines to the building. A chemical root inhibitor was applied, and the geomembrane, a 0.5-mm-thick polyvinyl chloride (PVC), was installed in the trench (Fig. 8-5). Pieces of woven polypropylene were placed over some of the larger and sharper cut roots to prevent them from puncturing the vertical geomembrane. The trench was backfilled with the excavated native material, and an airpowdered hammer was used for compaction. Wetting boxes were installed alongside the building's distressed sections to keep the foundation subgrade from drying.

A year later a composite plastic was used to create a vertical moisture barrier around a Summit Road building in Noble Park, Victoria. The one-story brick office building was suffering settlement in the front of the structure. This resulted from the combined impacts of trees in the garden and the reactive clays underneath the building (Fig. 8-6). A consultant employed to study the problem recommended that a reinforced-concrete cut-off wall be built. In a trial test, a PVC geomembrane mat was found to experience drying and cracking because of the nature of the soil, and it was therefore rejected as a vertical moisture barrier.

**Figure 8-5.** Dr. Holden with polyeurethene barrier.

**Figure 8-6.** Australian barrier and footpath.

Instead, it was decided to create a composite using PVC and a polyethylene, either LDPE or HDPE. A miniexcavator cut a trench 2 m (6.5 feet) deep and 210 mm (8 inches) wide. Water and root passageways around service pipes were prevented by the use of a polyurethane expanding-foam application. The expanding foam was found to bond effectively to the plastic barrier sheets, serving as a barrier, which was an economical solution. The backfilling material was a screening that provided a water reservoir for the foundation clays, saving the use of saturation boxes to provide the moisture for the rewetting of the foundation clays.

An experimental site was selected on Melrose Drive, Tullamarine, Victoria, in 1992 to determine the effectiveness of the moisture and root barriers in containing the drying results from the Sugar Gum and Angophoram trees found in the median. Excavating for the trench uncovered basaltic boulders, or floaters, up to 400 mm (15 inches) in width. The chain trencher was not able to handle the floaters, and a 300-mm (12-inch) bucket on a miniexcavator was required along with a hydraulic pick to deal with the larger rocks. A 1-mm, high-density polyethylene (HDPE) was used and backfilled with the excavated clay from the trench. A 300-mm (12-inch) Stanley vibrating plate was used to compact the backfill, which also received a root inhibitor. A polyurethane foam was used to provide a seal around a drainage pipe. The problem of dealing with basaltic boulders continues to challenge engineers.

In 1993, the St. Kilda City Council, now the Port Philip City Council, sought assistance in dealing with the distortion to a two-story block of apartments caused by two large Planer tree roots. It was decided to try to create a barrier using a polyurethane (PU) foam. A special formulation was developed of Trifluralin, a herbicide that inhibits root growth, and polyurethane. Prior testing found that the PU barrier, if untreated with the herbicide, could be penetrated by roots.

The work on Shelley Street, Elwood, involved the use of a new slim-line boom for the chain trencher (Fig. 8-7). It excavated a trench 70 mm ($2^{7}/_{8}$ inches) wide for a 1.8-m (5.8-foot) depth. The trench was cut in the footpath to maximize the distance between the trees and the building. Herbicide-treated PU foam barriers were constructed in panels that were 2 m (8.5 feet) long. Two inspection wells were drilled between the foam barrier and the building to facilitate later inspection to see if any roots had been able to penetrate the PU foam. After 1 year, significant differential movement took place across the barrier, creating an elevation difference of 60 mm ($2^{1}/_{2}$ inches) in the asphalt footpath. This gap created a potential safety hazard for footpath users. It was decided that in the future barriers would be built between the footpaths and the buildings. It was also determined that the main disadvantage of the PU barrier was that it was more expensive than the reinforced-concrete walls.

In mid-1993 the Melbourne City Council (MCC) requested assistance in dealing with expansive soil damage to residences on Drummond

**Figure 8-7.** New Australian slim-line trencher boom.

Street, Carlton. The structures were founded on basaltic clays containing large boulders. In this remediation work, the chain trencher was not used; instead, a miniexcavator was substituted to cut a trench that was 2 m (6.5 feet) in depth and 300 mm (12 inches) in width. A new product, Liquifill, a cementaceous flowable backfill material, was used as it had no compaction requirements, could completely fill any cavities, would bond to the subgrade, and would have a low permeability (Fig. 8-8). This flowable backfill was used against a 1-mm HDPE placed along the house side of the trench, and it was overlapped at each service pipe. Special steps were taken to prevent intrusion of the elm tree roots.

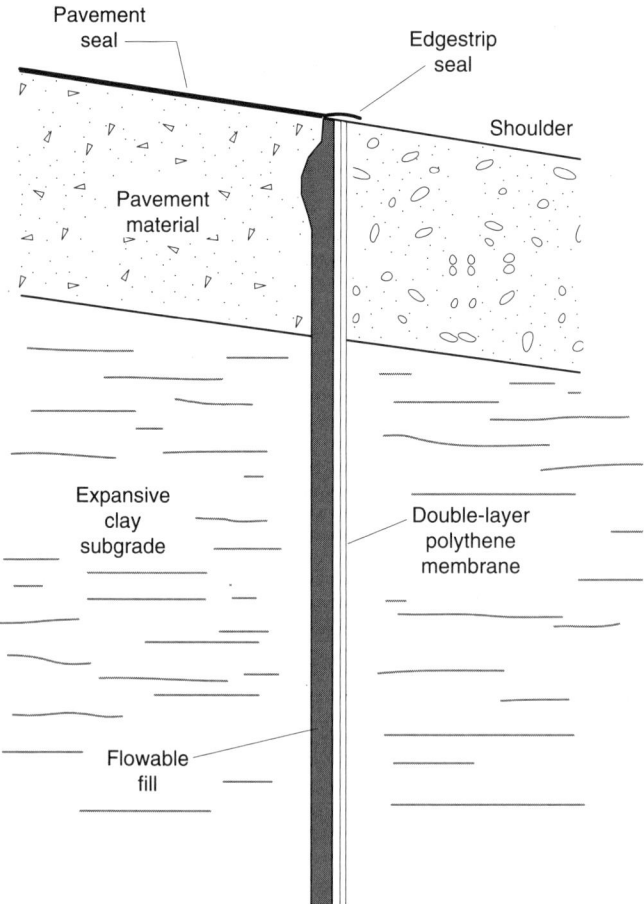

**Figure 8-8.** Australian flowable backfill trench.

Buildings 83

Again in 1993 the MCC sought assistance in dealing with tree roots that were causing problems to several terrace residences in the Wolseley Parade area of Kensington. The MCC was concerned that the stripping of the bark by the existing trencher would make the tree more prone to disease. To protect the tree, special saw teeth were designed and fitted to the slim-line trencher boom. This adaptation permitted the trencher to cut through boulders of up to 300 mm (12 inches) in diameter. A miniexcavator was used to excavate around the closely spaced house services. The 1-mm HDPE sheets were placed in the trench closest to the houses. A root inhibitor was applied with the Liquifill flowable backfill, which was delivered to the site in a transit mix truck. A special hopper was designed to pass the backfill from the truck to the trench. Liquifill was chosen for its ability to displace any water in a trench.

In mid-1993 another project was carried out for the St. Kilda City Council. A house on Raglan Street, St. Kilda East, was being distorted by roots from large Planer trees. This project was carried out similarly to past ones except for a modification at the service lines. This modification consisted of a ramp that was made for the trencher to pass over the service line. In discussing this project, Holden suggests a more effective manner of dealing with a 1-mm HDPE. He notes that placing an HDPE around house services is labor intensive, time-consuming, and expensive. An alternative method is being developed by VICRoads and the faculty of the Agriculture, Forestry and Horticulture Department of the University of Melbourne. This group is developing a waterproofing chemical that can be added to the Liquifill to make it relatively impermeable. With the additive, the Liquifill would be able to, by itself, seal any cracks that developed and thereby block any root intrusion. This work is being led by J. C. Holden and Kerry McManus, affiliated with VICRoads and Swinburne University, respectively, has tremendous potential for the future.

## 8.3 New Developments Using Geomembranes

The significance and importance of the work being pioneered by the Australian group—VICRoads, Swinburne University, and the various city councils in the state of Victoria—cannot be overemphasized. This group is addressing the challenge of the destructive movements of expansive soils aggressively and successfully. The benefits of their work will be felt all over the world.

In the time that has elapsed from the Harvard meeting in 1938 to the present, a tremendous amount of progress has been made in preventing expansive soil catastrophes. The Australian group alone has made significant contributions to the modern understanding of this hazard and its prevention. They have developed procedures for controlling expansive soils that plague so many building sites. One of their most significant achievements has been the development of a slim-line trencher, which can excavate 6 to 6.5 feet deep, opening a trench that is only 3 inches wide that can be filled using a flowable backfill.

## 8.4 References

1. W. E. Simpson, Sr. "Foundation Experiences with Clay in Texas." *Civil Engineering*, ASCE, November 1934.
2. D. Wooltorton. "A Preliminary Investigation into the Subject of Foundations in the Black Cotton and Kyatti Soils of the Mandalay District, Burma." *Proceedings of the International Conference on Soil Mechanics and Foundation Engineering*, vol. III. Harvard University, Cambridge, June 22–25, 1936.
3. Building Research Advisory Board, "Criteria for Selection and Design of Residential Slabs on Grade," Federal Housing Administration, Washington, D.C., 1968.
4. Fu Hua Chen. *Foundations on Expansive Soils.* Elsevier, Amsterdam, 1975.
5. W. G. Holtz and S. S. Hart. "Home Construction in Shrinking of Swelling Soils," Colorado Geological Survey, special publication 14, Denver, 1978.
6. J. C. Goode, D. J. Hamburg, and J. D. Nilson. *Moisture Content and Heave Beneath Slabs on Grade.* Fifth International Conference on Expansive Soils. Adelaide, South Australia, 1984.
7. P. C. Noe, C. L. Jochim, and W. P. Rogers. *A Guide to Swelling Soils for Colorado Homebuyers and Homeowners.* Colorado Geological Survey, Department of Natural Resources, Denver, 1997.
8. P. C. Noe and M. D. Dodson. *The Dipping Bedrock Overlay District.* Colorado Geological Survey, Department of Natural Resources, Denver, 1995.
9. R. L. Handy, "The Day the House Fell," ASCE Press, American Society of Civil Engineers, New York, 1995.
10. J. C. Holden. *Some Developments in Urban Root Barriers.* Seventh Australia-New Zealand Conference on Geomechanics, July 1996.

# 9
# Case Studies: Buildings without Geomembranes

Case studies about buildings where geomembranes have not been used are worthy of examination. This chapter profiles the class-action suits that have resulted in some instances, in which damage was sustained by unprotected buildings.

## 9.1 Class-Action Suits in California, Colorado, and Texas

Class-action suits in California, Colorado, and Texas have been filed by groups of home owners whose homes, service lines, streets, curbs, and sidewalks have been damaged. The claims indicate that adequate protection against the hazard of expansive soils was not ensured. In many of these cases the old-world rule of 'Omuerta,' or silence, seems to have been the order of the day. The terms of settlement in these lawsuits have been sealed in some cases, which has hindered the flow of information about expansive soil damage. The suits seldom enhance anyone's reputation, and they almost always impose severe financial penalties. Lawyers help their clients in these cases seek the "deep pockets." At one engineering meeting, a lawyer spoke at length explaining how to go after the "deep pockets." When the lawsuits reach the courtroom, often defendants have their own experts, frequently engineers, from the academic community testifying against the plaintiff's experts who are also frequently engineers from the academic community.

A March 1994 article in the *Contra Costa Times* reported on a settlement with a builder for $2 million. The claim of the home owners reported the shifting of lots and the cracking of homes, as well as numerous other problems. No mention is made of expansive soils in the article, though the area is one of several in California where expansive soils have been identified. The claimants have received additional funds for damages they claim are due to the soil movements, bringing their total home owner awards to more than $4 million.

According to a series of articles concerning the Denver, Colorado, metro area, a series of class-action suits have been filed by home owners who claim that the builders are liable for damages to their homes caused by expansive soils. Articles have appeared in the *Denver Post*, the *Wall Street Journal*, and the *Rocky Mountain News* recounting the events. Stories printed in 1996 indicate suits filed against Highlands Ranch, Richmond Homes, US Home, Merit Home, and Ryland Home. Falcon Homes settled a similar suit filed by 305 Highlands Ranch home owners for $2.5 million. A 1997 media article reported a court settlement of expansive soils damage claims by home owners in the Mission Viejo subdivision. The jury found for the home owners on four of the five counts claimed. The four counts were that the Mission Viejo's builders were negligent, they violated Colorado's disclosure laws on expansive soils, they breached implied-warranty laws, and they failed to comply with the state's consumer protection act. The jury found the defendant not guilty of being "willful and wanton" in their actions. The estimated cost of the damages could total $33 million according to a newspaper article.

Mission Viejo is the owner of the 22,000-acre Highland Ranch development. The company quit building homes there several years ago but continued to sell lots for homes and other developers. In the Highlands Ranch lawsuit filed by 937 home owners, it is claimed that the firm's own geotechnical engineers identified the high-swell-potential areas. The home owners further claim that they were not adequately informed. The defendant's attorney has indicated that they will appeal, and the Mission Viejo senior vice president has said that they have done everything to build the homes properly.

Rather than going to court, Richmond Homes, Colorado's largest home builder, settled four class-action suits. All related to damages it was claimed were caused by expansive soils. If the Douglas County District Court approves, this settlement would apply to 12,000 homes built by Richmond since 1988. The arrangement would provide extended warranties to the home owners, providing for repairs if necessary. The company has denied all allegations of negligence, but they will be responsible for the plaintiff's legal fees of $1 million. A key issue was the builder's use of concrete slab basement floors resting on expansive soils, which may have caused severe cracking and distortion of the basement floors and other home features.

Corpus Christi, Texas, is located on the coast of the Gulf of Mexico, approximately 2000 miles from Contra Costa County, California, and 1000 miles from the Denver metropolitan area in Colorado. A number of Corpus Christi home owners have filed law suits claiming that their houses were no longer habitable due to damages from expansive soils. One would tend to think the coastal city would be founded on sand. Not so. Rather, the geological makeup is similar to the Arabian peninsula, where a Saudi king's palace is suffering expansive soils damages. In Corpus Christi, with lawyers and university professors in attendance, claims have been made concerning missing steel in slab foundations and leaking service pipes that are wetting the clays. In any event, the houses have been cracking, doors have been jamming, and windows have been stuck.

Reputedly, after settling the Corpus Christi suits, the lawyers moved onto San Antonio, Texas, a city with well-known areas of expansive soils. San Antonio residents have been coping with these soils in numerous ways for decades. Willard Simpson, Sr., had worked in that city from the early 1900s seeking better methods to minimize damages from expansive clays. While working on a Maverick family home, Simpson developed the drilled-shaft foundation. Probably many of the factors identified in the Corpus Christi confrontation reappeared in San Antonio. This is true despite implementation of the BRAB and PTI design recommendations.

Fu Hua Chen in his book *Foundations on Expansive Soils* reviews many of the damages caused by these soils. Many of the 1990s suits in the Denver area share the same type of damages that Chen described in the 1977 edition of his book. One of the 1990s suits involved 1000 home owners. Experts have been assembled on both sides of the courtroom. Lawyers have received considerable newspaper coverage on their ongoing efforts to seek reimbursement for their clients, the damaged home owners.

## 9.2 Managing Expansive Soils around Large Institutional and Commercial Buildings

Large institutional and commercial buildings also face the challenge of dealing with expansive soil. The development by Simpson, a consulting engineer, and Ed Duderstat of San Antonio, Texas, of the drilled-shaft foundation in the 1920s established a preventive measure often followed to this date. In an area of expansive soils, the foundation for a large industrial or commercial building would have drilled shafts. The shafts would be founded on a bedrock or at least in a segment of moisture-stable material and would be drilled to an appropriate depth. A cage of reinforcing steel would be placed in the shaft, and the concrete poured. This has worked well usually, and it is used today in many locations. Frequently in the southwestern United States, a drilled shaft is the foundation of choice for buildings and bridges.

Interesting too is the fact that Simpson designed the foundation for the Smith Young building, now called the Transit Tower, in San Antonio, and he included some extra precautions. Simpson had jacks placed in the substructure in the eventuality that they were needed to restore the building to a plumb condition. Bell footings became an added feature of the drilled-shaft foundation, the concept being that the added area of concrete at the founding elevation would provide a bonus and an economic benefit. Several years after its introduction, some engineers became reluctant to use the bell because in some cases it apparently provided an additional area through which the expansive soil could push upwardly on the building.

Though the drilled-shaft foundation has many virtues, it does not come with a guarantee. A large institutional building in the San Antonio, Texas, area had drilled-shaft foundations founded below the zone of moisture activity, yet building distortions still occurred. Investigations revealed that the expansive soil, a swelling clay, had adhered to the shafts and had literally torn them in two. It appears that the moisture in the concrete found an attraction in the suction of the expansive soil, and its activity vented itself on the shaft. The lesson in this experience is that care should be taken to isolate the drilled shaft because it is founded below the zone of moisture activity.

Other studies, some led by Lyman Reese of the University of Texas, have investigated how much of the drilled shaft's load is actually transmitted to the founding elevation. In these studies it was found that skin friction of the shaft sides against the surrounding soil in many cases resulted in no load's reaching the shaft's footing. Instead, the skin friction accounted for the load placed upon it.

For those in Colorado, there's an added cautionary note. In Noe's research, shafts were not successful when founded in heaving or otherwise disturbed, steeply dipping bedrock [1]. The standard swelling soil values apparently are not satisfactory as they do not take into account the action of the rock blocks. This presents an additional consideration to be taken into account in similar geologic areas, although it may apply more to one- and two-story structures where a 20-foot drilled shaft meets the building's needs. Its applicability is worthy of analysis for buildings requiring deeper shafts too.

## 9.3 More Than 2000 U.S. Buildings Damaged by Expansive Soils

When Fu Hua Chen wrote his book in the 1970s, he noted that he had investigated over 1000 buildings in the Denver and Front Range areas of Colorado that had suffered damages caused by expansive soils. Judging from the class-action suits, the number for the United States should exceed twice the 1000 for the Denver area alone. It is difficult to gather precise data because many of the class-action suits there have been settled with their records sealed by the courts. A similar situation exists in Texas. First, apparently the lawsuits began in the Corpus Christi area of the state. The legal followthrough then moved onto San Antonio, about a 160 miles inland from Corpus Christi and the Gulf of Mexico coast. In 1997 a San Antonio consulting engineer commented that it seemed every consultant in the city was engaged either in prosecuting or defending insurance companies in these suits. The two methods that had been used to deal with the expansive soil were lime treatment administered through a pressure injection of a lime slurry or replacement of the swelling soil with a nonexpansive material.

The work of Kirby Meyer, an Austin, Texas, consultant, indicates that a considerable amount of damages have accrued to buildings in that area from expansive soils. In addition, Kent Wray's work in the Texas panhandle area indicates that more cases have been uncovered there. Still more cases have been discovered in north Texas, in California class-action suits in the San Francisco area, Contra Costa County, and in the valley just over the coastal range from San Diego.

## 9.4 Preventive Strategies

Remedies do exist for expansive soil problems for buildings under construction or already built. They vary from drilling shafts, to separating the grade beam from the piling or the shaft, removing the expansive soil that the grade beam is resting upon, and in some cases treating the soil by pressure injection of lime, lime fly ash, or a patented chemical mixture. In Australia's state of Victoria, numerous cases have been reported on the use of a vertical moisture barrier.

Geotechnical engineers and their partners in design and construction continue to search for ways to control the destructive movements of expansive soils. Their interest has been aided by the research engineers at the universities, manufacturers, and trade associations, and taken together, their studies have been professional and at the same time pragmatic.

In Arizona expansive soils have caused only slight impacts on the buildings in their two major cities, Phoenix and Tucson. Swell in the Phoenix area has been documented to be usually 2 to 3 percent. In the Tucson area expansive swell has been estimated to be between 3 and 5 percent. Problems arise between houses that are close together where an air-conditioning unit or a vegetative planting may be disrupting drainage in the limited space. One interesting challenge developed at a large parking lot at a truck stop. The truck washing facility provided water, and the resulting wastewater seeped through the cracks in the pavement, which sent the Chinle clay into motion.

Keeping water away from clay is a necessity to minimize the expansive soil's destructive movements. Concrete slabs and gutters may help minimize the problem there. Sometimes interior floor slabs are buckled in extreme cases.

The presence of the dipping-bedrock formations in the Tucson area have raised considerable concerns, and builders have consequently avoided the area. Between the two cities, there may be 50 to 100 claims against builders relating to expansive soils. The greater challenges from the Chinle clay may lie in the northern part of Arizona. Both the northeast and northwestern corners of the state reputedly have serious swelling soils problems. The Navajo Indian reservation is severely impacted, especially their school district buildings.

A widely respected professor and winner of the Terzaghi Award reports on some of the complexities involved in dealing with expansive soils and their impacts on buildings. In Victoria, Texas, expansive soils caused an expensive home to suffer substantial cracking and heaving even though the foundation had been carefully designed. Investigation revealed that the reinforcing steel was in place and that the concrete met all the strength requirements. To date, no plausible explanation has been found.

However, the professor who led the investigation pointed out some possible contributing factors. He listed port tensioning cables found lying loose in the bottom of the slab on grade, which means that there has been no tensioning, post or otherwise. In addition, he has found no steel at all in the slab designed to be a steel-reinforced foundation.

South Africa's engineers have also sought solutions to the destructive movements of expansive soils. Flooding a proposed building site to pre-heave the structure is one way they have addressed the problem. They note that the technique is successful only if it is properly carried out [2].

Australia and Canada have reported on their substantial efforts to control damages related to swelling soils. At Swinburne University of Technology in Melbourne, Australia, investigation focused on developing a design formula for slabs to be built on expansive soils. They found that the PTI and Walsh methods yielded results close to their formulation, which they called appropriately "the Swinburne method." They also concluded that the BRAB design was overly conservative [3].

More than 50 years ago, Wooltorton investigated extensive damages to the buildings in the Mandalay District of Burma. He found that it wasn't design or construction workmanship failures that were responsible. Rather, the cause was expansive soils. His studies were the basis of many recommended preventive strategics. For example, drainage systems should divert water from the buildings. In addition, there should be no planting around the buildings. At a recent Texas Section ASCE meeting, two papers were presented that addressed this question. One of the papers looks at trees that are no longer living but whose roots continue to have an impact on the site [4].

Other preventive strategies have proven successful. For example, flower beds should be away from the walls of the building. Also, geomembranes can isolate water to keep it from penetrating a foundation subgrade. Trees should be sited far enough away from a building to prevent the drip line of the fully grown tree from intruding into the foundation. These considerations are applicable in any situation in any country where expansive soil is present.

## 9.5  References

1. Fu Hua Chen, "Foundations on Expansive Soils," Elsevier, Amsterdam, 1975.
2. D. C. Noe. *A Pierre Shale Primer*. Prepared for the Douglas County Planning and Community Development, Colorado Geological Survey, Denver, 1994.
3. D. C. Noe, C. L. Jochim, and W. P. Rogers. "A Guide to Swelling Soils for Colorado Homebuyers and Home Owners," Colorado Geological Survey, Denver, 1997.
4. W. K. Wray, Editor. "So Your Home Is Built on Expansive Soils," American Society of Civil Engineers, New York, 1995.
5. J. C. Holden, "Some Developments in Urban Root Barriers," The Australia New Zealand Conference on Geomechanics, Adelaide, Australia, 1996.
6. J. E. Holland, W. G. Pitt, C. T. Lawrence, and D. G. Cumind. *The Behaviour*

*and Design of Housing Slabs on Expansive Clays.* Fourth International Conference on Expansive Soils, Denver, 1980.

7. G. C. Blight, K. Schwartz, H. Weber, and B. L. Wild. *Preheaving of Expansive Soils by Flooding—Failures and Successes.* Seventh International Conference on Expansive Soils, Dallas, 1992.

8. R. V. Reed and J. Phipps. "Post Construction Effects of Preexisting Trees." *Proceedings of the Texas Section ASCE,* September 1996.

# 10
# Case Studies: Texas Highways, Roads, and Streets

This chapter continues to get down to cases. Its focus is on highway, roads and streets. It presents the what, where, when, and how. It reviews the early projects in Colorado and South Dakota where geomembranes were used. Texas projects and over 50 projects in Wyoming are examined. Additionally geomembranes on almost a dozen other state's highways are reviewed as are those in Australia and Israel. You'll leave the chapter knowing you can place the geomembrane safely and economically.

## 10.1 Lessons Learned from Projects in Colorado and South Dakota

Early projects in Colorado and South Dakota using geomembranes followed extensive efforts using other methods to control expansive soils. In one trial in 1933, Texas highway departments installed asphalt membranes horizontally in efforts to control moisture changes, which were leading to volumetric changes in expansive clays. Later, in the 1950s, W. J. Van London, the Texas Highway Department expressway engineer in Houston, used heavy asphalt encapsulation on several embankments [1].

Conard Kelly, a district laboratory engineer for the Texas Highway Department, had used lime in expansive clay subgrades while he was in federal service during World War II. The procedure he developed then and later during his work for the National Lime Association became widely used in Texas, Oklahoma, and New Mexico. The treatment involved determining the subgrade's Atterberg limits, the unit weight of the subgrade, the amount of lime to be added depending on the PI, and the density of the material. The higher the PI, the greater the percent of lime to be added. Generally the top 6 inches were treated. This helped, but with passing years, it was frequently considered to be an inadequate treatment depth. Consequently, the department increased treatment depths on highway work up to 24 inches. Belgium in one case treated the top 5 feet. Despite the increase to 24 inches in some projects, these greater depths have not been wholly sufficient to control the expansive soil movements. Nevertheless, the state has continued to use lime treatments even when other methods have been recommended.

B. A. Brakey, Colorado Highway Department research engineer, had used heavy asphalt treatment of the expansive soils subgrade in efforts to control their swelling movements by minimizing their moisture changes. Colorado had previously, in the 1950s and 1960s, used lime treatment on these expansive soils. They also used drilled shafts with lime backfills and lime slurries in similar efforts. Brakey's work with the heavy asphalt applications of 1 gallon per square yard placed the treatment from ditch backslope to ditch backslope. The Du Pont Company had been observing Colorado's efforts. They concluded that the geomembrane they had developed, called Typar, could do the job more effectively and economically, so they offered a geomembrane for a trial section (Fig. 10-1).

The Typar was placed between two 6-inch layers of sand. Initial reports were enthusiastic. Brakey reported on its use at an Expansive Soils Technical Advisory Group (ESTAG) meeting held at the Waterways Experiment Station, which has been maintained by the U.S. Army COE as part of the Federal Highway Administration study of expansive soils. Hopes were high, and it led to the first use of Typar on a Texas Highway Department (now TexDOT) project. As a postscript,

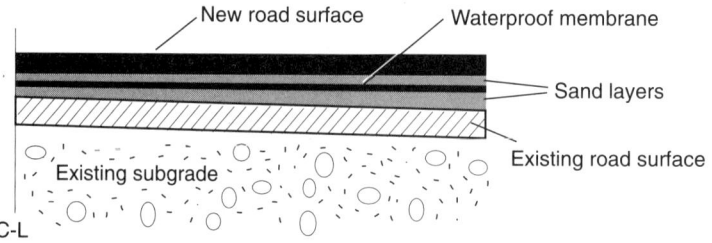

**Figure 10-1.** The first Colorado geomembrane trial.

several years later the Colorado DOT found that the sand layers both above and below the Typar were saturated. They concluded that the project results had not been satisfactory.

Eugene B. McDonald, South Dakota Highway Department materials engineer, reported at an ESTAG meeting on the use of a geomembrane as a vertical moisture barrier placed 4 feet deep along the shoulders of U.S. 12 in Corson County. The trial sections, one in a cut and the other in a fill, were in an area of expansive Pierre shale subgrade, which had resulted in severe pavement heaves. McDonald indicated that frost heaves are a factor to be considered. The trial locations had lime-treated subgrade with RC 1 asphalt and lime surface seal. A polyethylene plastic blanket was placed 4 feet deep along the shoulders from Stations 155 to 162, and 186 to 192. Work on the project took place in 1965. Its location in the Missouri River trench area was chosen because it suffers the most heaves of any section in the state.

Moisture tests were taken at these two sites and at the control sections between 1965 to 1971. Road roughness was also measured during this period. The moisture results and the roughness index indicated little difference between the geomembrane-protected sections and the control sections. McDonald's report concludes that there would have been little differences in moisture and roughness over the 5-year period but if the geomembrane had been placed twice as deep (8 feet), the moisture and pavement heaves would have been eliminated. McDonald also reported that the department had used lime treatment of the subgrade, but it was felt that the geomembrane offered the better solution [2].

## 10.2 Texas Highway Swelling Soil Projects

Texas highway swelling soil projects and their study can be traced to the 1930s. The studies in Navarro and Guadalupe counties in south-central Texas were reported at the Harvard conference in 1936 by Henry C. Porter, engineer of research, State Highway Department, Austin, Texas. Clay samples from the highway subgrades had liquid limits from 80 to 100, plasticity indices from 52 to 74, and shrinkage limits from 8 to 11. The samples were taken from a pavement laid in 1931 where the subgrade had cracks 6 inches wide and 10 feet deep [3]. Of interest are the moisture contents reported between 1934 and 1936 where major variations took place in the top 10 feet but were significantly less below that level.

In a later report Porter presented test results on the bearing strength of clays relating to density and moisture contents (Fig. 10-2). Included

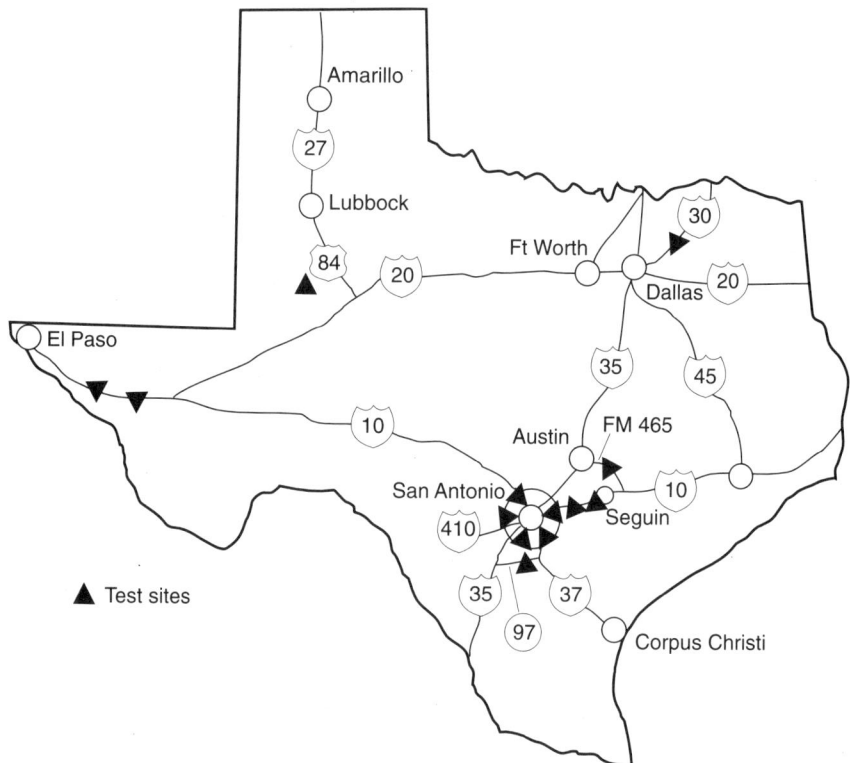

**Figure 10-2.** Texas geomembrane sites.

in this Guadalupe County Research Project were measurements in changes in subgrade and pavement elevations. Thirty-seven test sections, each 1200 feet long, provided the sample sources. Plastic indices generally were in the 40s and 50s, and moisture contents varied as the PIs increased. Tar paper was used on several of the sections to prevent the water in the concrete being drawn into the clay subgrade. The study showed the moisture change in the subgrade varying to 5 and 6 percent, which was too small to cause significant loss of strength [4].

Of interest is Porter's work on the Guadalupe Research Project, formerly U.S. 90, now IH 10, between San Antonio and Seguin. Atterberg limits were determined for sections of the pavement subgrade and related to distortion on the roadway surface. Not mentioned in this report was the fact that ponding was used in construction efforts to control the expansive clays.

In the 1960s Chester McDowell, with the Texas Highway Department, used the ponding technique on IH 35 sections in the Waco area. The project was ponded on separate sections usually below natural ground in a cut. A postconstruction study indicated that the ponded sections required the least amounts of overlay placed on their pavements, and the nonponded sections required the most [5].

Paul Teng, at the time a Mississippi Highway Department research engineer, now with the Federal Highway Administration, used the ponding method with apparent success on an interstate highway in the Jackson area. Teng had holes drilled 20 feet deep on 5-foot centers which were then filled with a pervious material prior to ponding. The ponding covered the section with water for 90 days. To date, it appears that these ponded sections are the only ones on the interstate system in the Jackson area that do not require rehabilitation.

The Texas Highway Department ponded sections of U.S. 90 in western San Antonio. On this construction, generally in the cut sections, earthen dikes contained the ponded water. The ponds were maintained for 30 days, after which they were drained. The subgrade then received a 6-inch treatment of lime, foundation and flexible-base courses, and a hot-mix asphaltic concrete pavement. Postconstruction testing indicated that the zone of moisture activity ranged from 6 to 8 feet below the surface. This information was later applied to the decision on how deep to place the geomembrane as a vertical fabric moisture barrier [6].

Beginning after World War II and continuing through the present, lime stabilization has been used extensively around the state to treat expansive soil subgrades. In several cases where deep, vertical fabric moisture geomembrane barriers were installed on rehabilitation projects, the subgrade had been lime stabilized earlier during the initial construction. This lime process was refined under Chester McDowell, a Texas Highway Department research engineer. Atterberg limits of the subgrade are used to determine the PI, and the amounts of lime to be added are increased with higher PIs and subgrade densities. Usually a 6-inch subgrade stabilization treatment is used. The lime is frequently delivered to the subgrade from a truck transport. It is then mixed using a portable self-propelled mixer and laid in with a maintainer. Water is added from a tank truck, after which, the mixture is compacted with rollers.

The first Texas project to use geomembranes was constructed in 1976. The Du Pont Company offered their Typar geomembrane to the Texas Highway Department to try out. The THD accepted Du Pont's offer and proceeded to prepare the plans and then award a contract for the work. They supervised the construction for the rebuilding of General McMullen Drive, a San Antonio urban arterial roadway (Fig. 10-3). Typar was originally developed in Europe. It is a spun-bonded polypropylene coated with an ethyl vinyl acetate (EVA), and it comes in rolls 3 m (9.75 feet) wide (Table 10-1). Before its trial use by the THD, Typar was subjected to significant preconstruction testing. In addition

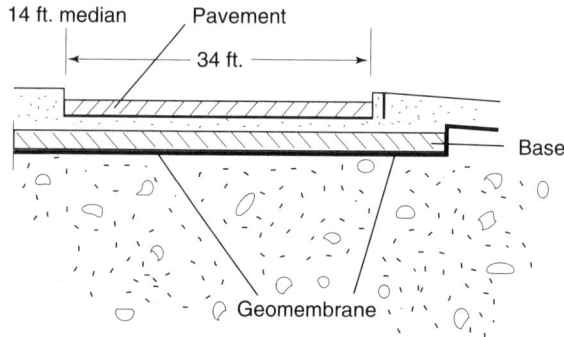

**Figure 10-3.** General McMullen Drive section.

**Table 10-1.** Physical Properties of Typar Styles 3153 and 3353, Spun-Bonded Polypropylene

|  | Style 3153 | Style 3353 |
|---|---|---|
| Weight, oz/yd$^2$ | 2.4–2.6 | 4.8–5.1 |
| Thickness, mils | 7–9 | 17–20 |
| Grab tensile, lb | 85 | 130 |
| Grab elongation, 5 percent | 67 | 55 |
| Trapezoidal tear, lb | 30 | 60 |
| Mullen burst, lb/in$^2$ | 70 | 150 |
| Permeability, perms | 3 | 25 |
| Puncture, lb/in$^2$ | 52 | 130 |

Specific gravity = 0.95
Melting point = 340°F

to Atterberg limits, its serviceability indices (Table 10-2), stiffness coefficients, maximum deflections, spreadability indices (Table 10-3), potential vertical rise (PVR), and photologging (Table 10-4) had been determined (Fig. 10-4).

It was decided to place the Typar in the areas of the highest PVR. The six-lane arterial street was constructed in half-width sections at a time. The half section was cut to grade, and the 600-foot test length was ready for the Typar. The Typar was unrolled horizontally across the subgrade by state maintenance personnel because the geomembrane was not included when the plans were prepared. This meant that the work was not sent out for bid. Base was dumped on the geomembrane test section

**Table 10-2.** Serviceability Indices for General McMullen Drive

|  | Outside ||| Center ||| Inside |||
| --- | --- | --- | --- | --- | --- | --- | --- | --- | --- |
|  | 7-81 | 12-81 | 6-82 | 7-81 | 12-81 | 6-82 | 7-81 | 12-81 | 6-82 |
| Northbound Lanes ||||||||||
| South control | 2.75 | 1.73 | 1.95 | 3.67 | 3.54 | 3.25 | 3.85 | 4.11 | 4.13 |
| Fabric section | 3.63 | 3.62 | 3.58 | 3.86 | 3.72 | 3.59 | 3.41 | 3.30 | 3.33 |
| North control | 2.17 | 2.81 | 2.92 | 3.69 | 3.68 | 3.67 | 4.14 | 3.88 | 3.87 |
| Southbound Lanes ||||||||||
| South control | 2.32 | 1.95 | 2.16 | 3.09 | 3.01 | 2.94 | 3.41 | 3.34 | 3.31 |
| Fabric section | 2.64 | 2.58 | 2.61 | 3.83 | 3.81 | 3.84 | 3.86 | 3.84 | 3.75 |
| North control | — | 2.70 | 2.60 | 3.68 | 3.61 | 3.64 | 3.61 | 3.53 | 3.56 |

**Table 10-3.** Spreadability Indices for General McMullen Drive

|  | Dates ||||||
| --- | --- | --- | --- | --- | --- | --- |
|  | 1/81 | 3/82 | 1/81 | 3/82 | 1/81 | 3/82 |
| Southbound Lanes |||||||
|  | L || M || N ||
| Control | 81 | 74 | 79 | 79 | 79 | 80 |
| Fabric | 86 | 78 | 83 | 83 | 83 | 83 |
| Control | 81 | 78 | 79 | 79 | 74 | 80 |
| Northbound Lanes |||||||
|  | T || S || R ||
| Control | 79 | 79 | 79 | 80 | 80 | 76 |
| Fabric | 84 | 82 | 84 | 82 | 81 | 77 |
| Control | 80 | 81 | 82 | 81 | 80 | 73 |

from trailer trucks, spread by maintainers, watered, and rolled; construction proceeded in stages for the full width of the roadway section. The Du Pont Company supplied two types of Typar for the project, Styles 3353 and 3153. The latter was used primarily along the outside edge of the northbound lane. The Typar 3353 weighed 5 oz/yd$^2$, and the 3153 weighed 2.5 oz/yd$^2$ (140 and 70 g/0.8 m$^2$, respectively). Though there was some movement of the geomembrane during the base dump-

**Table 10-4.** Photologging for General McMullen Drive

| Date | Section | L | M | N | R* | Average |
|---|---|---|---|---|---|---|
| 1/81 | Control | 1.08 | 0.97 | .51 | 0.98 | 0.99 |
|  | Fabric | 1.07 | 0 | 0 | 0.36 | 0.24 |
|  | Control | 0 | 1.05 | .69 | 0.54 | 0.36 |
| 6/81 | Control | 1.26 | 1.60 | 0 | 1.40 | 0.72 |
|  | Fabric | 1.42 | 0 | 0 | 0.50 | 0.32 |
|  | Control | 0 | 1.80 | .10 | 0.70 | 0.43 |
| 1/82 | Control | 0.63 | 1.25 | 0 | 1.64 | 0.59 |
|  | Fabric | 0.73 | 0.03 | 0 | 0 | 0.21 |
|  | Control | 0 | 1.43 | 0 | 0.16 | 0.27 |
| 8/82 | Control | 1.09 | 1.65 | 0 | 3.93 | 1.11 |
|  | Fabric | 1.64 | 0 | 0 | 0.53 | 0.36 |
|  | Control | 2.07 | 0 | 0 | 0.40 | 0.41 |

*Lanes S and T recorded 0 in all photologging throughout and were calculated as such in the average determinations.

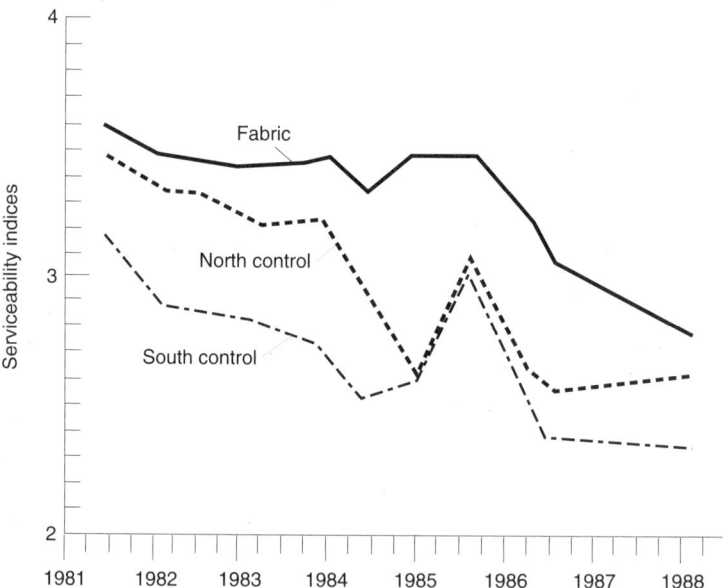

**Figure 10-4.** General McMullen Drive serviceability indices.

ing and laying operations, there was no sign of tearing. The first geomembrane placement was in June 1976, and the second was in June 1978. The typical McMullen section had three courses of asphalt-stabilized base, totaling 11 inches, a tack coat, and two ¾-inch hot-mix asphaltic concrete courses, which provided the riding surface.

Following construction, extensive testing took place. Elevation readings were taken on the pavement in the control and geomembrane sections. During the test period from 1977 to 1981, the changes were minimal with a maximum of 0.15 feet in the control section. The elevation changes in the geomembrane area were a maximum of 0.10 feet, compared to the PVR of 0.37 feet (4½ inch). Photologging was conducted between 1981 and 1982. Pictures were taken by a camera mounted 8 feet high on a moving vehicle. The pictures were projected onto a screen with 100 grid squares, and the cracks were tabulated. The highest average reading was 2.07 percent in the control section. In all cases it was found that the geomembrane section had less cracking than the control sections.

Dynaflect testing was conducted in 1973, prior to the 1981 and 1982 construction, using five geophones. Stiffness coefficients, spreadability, and maximum deflections were calculated. The stiffness coefficients separate the subgrade values from the pavement section. The higher the number, the greater the estimated strength. Preconstruction testing indicated that the subgrade in the geomembrane-protected section had lower values than the adjacent control sections. Following construction with the geomembrane in place, the subgrade strength was higher than the control sections. Spreadability indices indicating total pavement strength showed the geomembrane section to have higher values than the control sections.

Serviceability indices were determined using a profilometer. The profilometer readings were computer reduced to provide the serviceability index (SI). The perfect smooth surface received the highest reading, a 5. Descending readings indicated decreasing smoothness, a rougher surface. Dynaflect testing determined the spreadability index, and moisture sensors were installed. Photologging was conducted, and surface crack surveys were developed. Much of the testing continued through 1985. Surface inspection indicated that no patching had been required on the geomembrane sections. The indications from all the tests were that the geomembrane section provided a smoother ride with higher serviceability indices, higher spreadability indices, and less surface cracking. The moisture sensors didn't survive for a significant period and therefore provided no substantial data. The geomembrane added strength to the section, minimized pavement movement, and had less pavement cracking. The project that had previously required pavement repairs frequently required none for a decade [7].

The results from the General McMullen Drive geomembrane project were so encouraging that the Typar geomembrane was included in a

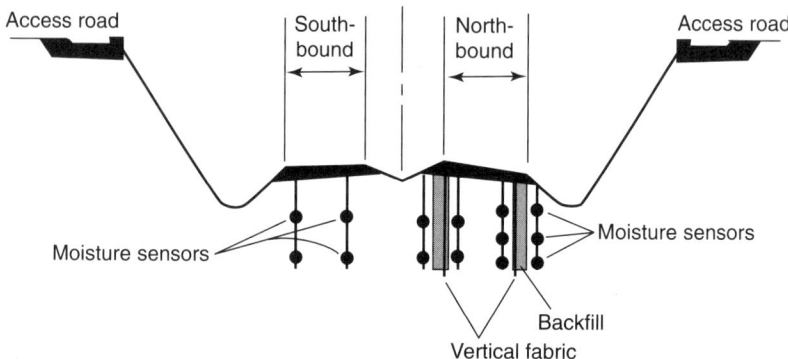

**Figure 10-5.** IH 410 typical section.

half-mile test section of the 1978 IH 410 rehabilitation project in southwest San Antonio (Fig. 10-5). This was the first Texas highway project in which a deep vertical fabric moisture barrier (DVFMB) was used. IH 410 is a four-lane divided rural freeway. Built in 1960, its section included 6 inches of lime-treated subgrade, 16 inches of foundation course, 8 inches of flexible base, a 3-inch lift of Type A asphaltic concrete, and 2 inches of a finer-graded, Type C asphaltic concrete riding surface. Each main lane had two driving lanes of 12 feet with a 10-foot outside and a 4-foot inside paved shoulder. A 44-foot sodded median separated the north and southbound lanes. Subgrade Atterberg limits had PIs ranging from 28 to 48.

In the Valley Hi Drive underpass area, the hot-mix asphaltic concrete pavement had suffered severe distortion with base exposed despite constant maintenance. The design decision on the 15-mile rehabilitation project was to place a geomembrane 8 feet deep, tacked 2 feet to each northbound lane shoulder for one-half mile through the underpass. The 8-foot depth was partially a result of the department's ponding project on U.S. 90 West, indicating a zone of activity in the subgrade where moisture changes were most pronounced. On the Loop 410 project, the adjacent southbound lane would receive only the level up- and asphaltic hot-mix concrete overlay serving in the underpass area as the control section.

The contractor bid the DVFMB at $20 per foot and began to excavate the geomembrane trench with a small tractor-mounted backhoe. Soon after the excavation began, the trench wall slid, and the contractor told the district engineer that the work couldn't be done. The district engineer reputedly replied, "You bid it, you'll build it." The contractor then hired a subcontractor who specialized in utility work. He used a larger, crawler type of John Deere 690 B backhoe with a sliding shoring attachment. The work proceeded, and the geomembrane, a Du Pont Typar T

063, was placed vertically 8 feet deep, tacked with an emulsion to the shoulder. The trench was backfilled as specified with sand. The maximum days' production of DVFMB was 400 feet.

Postconstruction testing included using the profilometer to determine the serviceability index (Table 10-5), photologging for crack survey studies (Table 10-6), and moisture sensors (Fig. 10-6). In each case the northbound lane with the geomembranes had the smoother ride, a higher SI than the southbound lane, and fewer surface cracks than the southbound lane. Eighteen moisture sensors were placed inside and outside the northbound lanes' DVFMBs and outside the southbound lanes shoulders. They did not last long enough to provide substantial information, but while they survived, they indicated less moisture variation inside the geomembrane-protected area [8].

**Table 10-5.** Serviceability Indices for IH 410

|  | Southbound lane, control | | Northbound lane, vertical barrier | |
|---|---|---|---|---|
| Date | Outside | Inside | Outside | Inside |
| 6/79 | 4.13 | 4.02 | 4.16 | 4.11 |
| 8/80 | 3.19 | 3.30 | 3.83 | 3.83 |
| 7/81 | 3.43 | 3.29 | 3.74 | 3.61 |
| 6/82 | 3.47 | 3.25 | 3.66 | 3.67 |
| 9/83 | 3.15 | 2.89 | 3.47 | 3.50 |
| 9/84 | 2.99 | 2.81 | 3.47 | 3.54 |
| 4/85 | 3.03 | 2.62 | 3.65 | 3.61 |

*Note:* A new profilometer was used from July 1981 forward. Both lanes were rotomilled April 1981. Both lanes were leveled up. There was twice the tonnage on the southbound lane than on the northbound lane.

**Table 10-6.** Photologging for IH 410

|  | 8/80 | 12/80 | 8/81 | 1/82 | 8/82 | 1/83 |
|---|---|---|---|---|---|---|
| | | | Northbound Lanes | | | |
| Outside | 0.07 | 0.08 | 0 | 0.04 | 0 | 0.19 |
| Inside | 0.01 | 0.24 | 0.08 | 0.34 | 0.03 | 0.37 |
| | | | Southbound Lanes | | | |
| Outside | 0.28 | 0.62 | 0.11 | 0.87 | 0.29 | 0.75 |
| Inside | 0.24 | 1.01 | 0.10 | 0.78 | 0.51 | 0.45 |

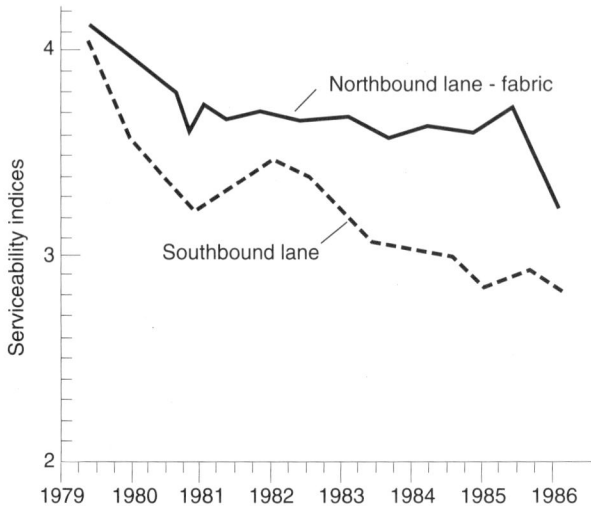

**Figure 10-6.** IH 410 serviceability indices.

A month after placement of the geomembrane with the sand backfill, a vehicle strayed off the shoulder and got stuck in the sand. A few weeks later another vehicle did the same thing. Concerns were raised, and it was suggested that cement be mixed with the sand, but the suggestion was rejected. No further reports of vehicles' getting stuck in sand were filed, and the cement stabilizing operation didn't take place. Four years after project completion, some pavement maintenance was done in the underpass area. Twice as much asphalt level-up was placed on the southbound lane compared to the northbound DVFMB lane. A decade or more later, the lanes on IH 410 were widened. Both the northbound and southbound lanes in the underpass area received a DVFMB.

The next DVFMB project awarded by TexDOT was on IH 37 in southeast San Antonio (Fig. 10-7). The construction of the eight-lane urban

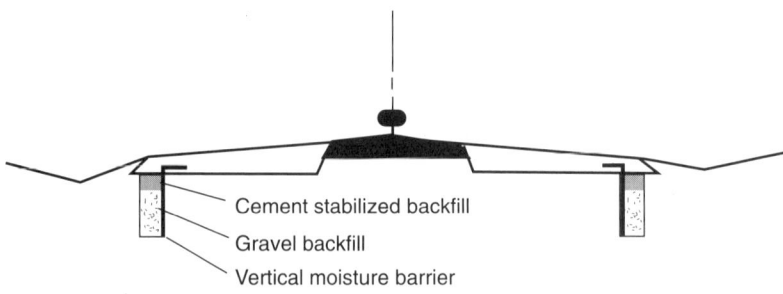

**Figure 10-7.** IH 37 typical section.

freeway was completed in 1968. A 2-mile section, from Hackberry Street south to Pecan Valley Drive, had 12-foot driving lanes, a sodded median varying from 24 to 36 feet width, and a 3-foot centerline concrete ditch with a steel beam barrier fence. The 2-mile section is generally 20 feet below natural ground in a Houston black clay series with PIs up to 54. Original construction had 6 inches of lime-stabilized subgrade, 8 inches of cement-stabilized base, and an 8-inch concrete pavement. In the 12 years following construction prior to the rehabilitation contract, the 2-mile section required asphaltic level-ups once or twice a year. Cost frequently exceeded $100,000 annually. When the district engineer visited with the local state legislator to explain the new level-up planned for the coming year, the response was not enthusiastic. The legislator reminded him that they had been doing that every year. Thinking quickly, the district engineer replied that this year they were doing something different. "We're adding a deep vertical fabric moisture barrier."

Prerehabilitation construction testing took place, including profilometer and moisture sensor readings. The roadway sections were then redesigned to drain from the median centerline to the outside shoulder ditches. The DVFMB chosen is 8 feet deep, and it would be placed along the outside shoulders. An underdrain was to be placed in the bottom of the trench, and a gravel backfill was specified with an 18-inch cement-stabilized trench cap. The rehabilitation contract awarded in 1980 also called for asphaltic concrete level-ups and finish courses. The low bidder contractor's price for the 21,485 square yards of geomembrane DVFMB was $21 per square yard of Du Pont's Typar T 063. The same subcontractor who did the DVFMB on Loop 410 was selected to excavate the trench and place the fabric. This time the subcontractor chose to use a Parsons 500 trenching machine to do the excavation with a special attachment to deliver the spoil material to waiting trucks. When the daily goal for placing the geomembrane of 400 feet was reached, not infrequently around noon, the operation shut down until the next day. A portable rubber-tired device pulled by the trencher held the geomembrane roll before turning it down to the excavated trench. The cement-stabilized base used as a cap for the trench was mixed in a portable concrete batching vehicle equipped with a chute to pour the mixture onto the gravel backfill.

Postconstruction testing with the profilometer gave SI readings for the 2-mile geomembrane and the control sections to the north and south (Fig. 10-8). Initial results indicated that the rehabilitated project with the DVFMB had higher SI readings and a smoother ride. Following an asphalt level-up of the north control section, it had the higher SI readings. The early pavement crack survey indicated less pavement movement and cracking on the DVFMB segments. Moisture sensor readings initially indicated less moisture fluctuation inside the geomembrane-

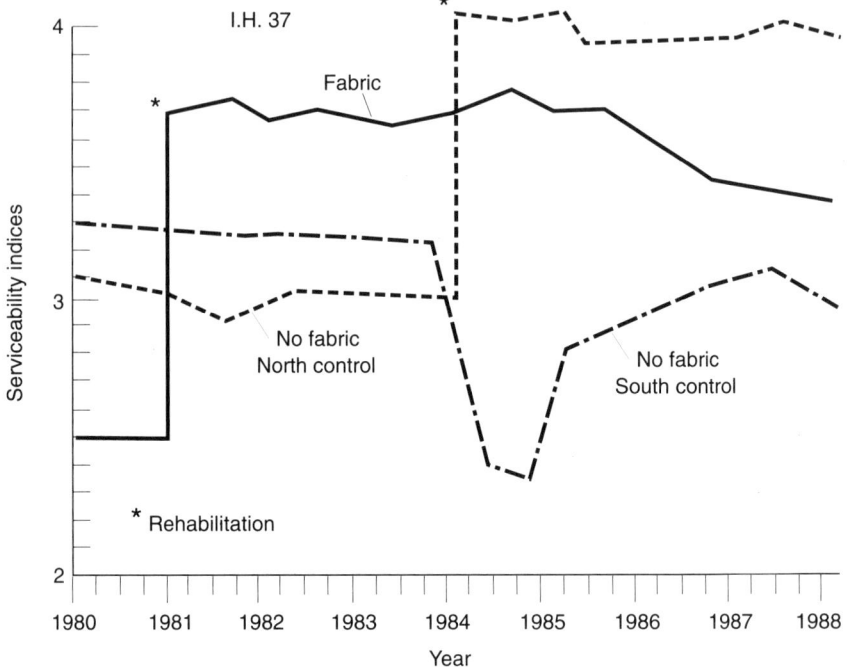

**Figure 10-8.** IH 37 serviceability indices.

protected subgrade. Instead of the semiannual or annual level-ups usually required, no maintenance pavement work was necessary for 12 years.

In 1983 two additional projects were awarded for which DVFMB was to be used. One on U.S. 281, the McAllister Freeway, is located in north-central San Antonio, just south of Loop IH 410. A multilane urban freeway, the rehabilitation of the southbound lane varied in grade from natural ground to a 20-foot cut section. The pavement suffered significant distortion thought to be caused by the subgrade whose PIs varied from 25 to 58. The design of the section constructed between 1970 and 1975 included 6 inches of lime-stabilized subgrade, 6 inches of base, an asphalt seal coat, and 8 inches of continuously reinforced concrete pavement. The southbound lane had suffered the pavement roughness problem and had received an asphalt level-up, an asphalt seal coat, and a hot-mix asphaltic concrete finish course. In addition, the DVFMB was positioned 8 feet deep along both shoulders with a trench gravel backfill and an 18-inch cement-stabilized base cap.

The contractor selected a Mirafi MCF 500 geomembrane, a 20-mil thick, polypropylene fabric with a bonded polypropylene film. A

Gradall backhoe was used for the trench excavation, and daily production averaged 200 to 250 feet. The geomembrane was unrolled on the pavement and manually turned into the trench. When progress seemed slower than desired, the contractor brought an additional Gradall to the project. Early profilometer testing was encouraging. The DVFMB southbound lane sections had the higher SI readings compared to the northbound control sections. Also, the sloughing trench wall problem seen on Loop 410 did not occur as the subgrade on U.S. 281 was stiff enough to pose no sliding complications.

The second of the two 1983 TexDOT projects using geomembranes was located on IH 10 on San Antonio's east side (Fig. 10-9). This rehabilitation contract extended from Pine to Amanda streets. During its initial construction in 1968, difficulties arose from springs when the grade line was in a depressed section. The subgrade was a Houston black clay with PIs ranging from 35 to 55. The six-lane urban freeway had a sodded median with its main lane sections constructed on 6 inches of lime-stabilized subgrade, 6 inches of lime-stabilized base, an asphalt seal coat, and an 8-inch continuously reinforced concrete pavement. The section required significant and costly maintenance due to the swelling soil, which caused pavement distortions.

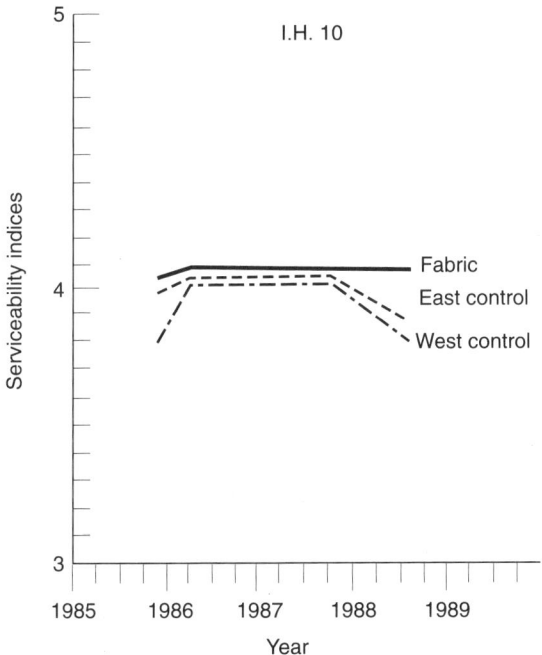

**Figure 10-9.** IH 10, San Antonio, serviceability indices.

The rehabilitation contract plans called for an asphalt rubber seal, asphaltic concrete level-ups, and finish surface as well as a DVFMB membrane to be placed along the shoulders of the eastbound and westbound main lanes. The geomembrane's trench was to be backfilled with Class B gravel and an 18-inch cement-stabilized base cap. As called for in the U.S. 281 rehabilitation contract, a generic geomembrane specification was used. The contractor chose a Mirafi MCF 500 fabric for the estimated 24,745 square yards of DVFMB bid at $15 per square yard. A trenching machine was used for the DVFMB excavation with a boom to place the material in waiting trucks. A sliding shoring was pulled by the trencher to hold and turn the geomembrane into the hole. Gravel was backfilled into the trench, and the cement-stabilized cap was delivered by transit mix trucks.

During construction, the trench wall began to slide. The solution adopted was to place the 8-foot trench 6 to 8 feet beyond the shoulder. In those areas the geomembrane was placed vertically 8 feet deep, then horizontally on the subgrade extending to a 2-foot tie to the paved shoulder. The horizontal geomembrane section was covered with 6 inches of subgrade as protection from the sun's ultraviolet rays. Despite this complication, the placement of the DVFMB exceeded 700 feet on several days. Profilometer testing followed completion of the rehabilitation contract. The geomembrane-protected sections provided the higher SIs and offered the smoother riding pavement [9]. The resident engineer noted that no maintenance work was needed on the section and that the DVFMB had made a believer of him [10].

In 1984 a test project contract was awarded for IH 30 near Greenville in northeast Texas. A rural four-lane divided freeway built over an expansive clay subgrade was experiencing considerable pavement distortions. Four test sections were established on the eastbound main lane with the adjacent westbound main lane serving as the control. The test sections were each 1000 feet long. In one section, the geomembrane vertical moisture barrier was placed 6 feet deep along both shoulders. In another, the geomembrane was placed 8 feet deep. Lime and lime fly ash slurry sections followed. The slurries were placed 8 feet deep in three staggered rows, parallel to the centerline, spaced at 1-foot intervals, along both shoulders of the 1000-foot lengths. The fabric sections used native material for the trench backfill with one sack-cement-stabilized cap 2 feet thick. The daily placement rate averaged 1000 feet. The contractor bid on the geomembrane, the excavation, and the cement cap at $27 per centerline foot and the lime and lime fly ash slurries at $12.73. Postconstruction testing indicated that the control sections had higher SIs and smoother riding sections than the geomembrane-protected sections. The initial reports also indicated that the geomembrane sections had more moisture changes than the lime sections. The project report indicated that the lime slurry and lime fly ash slurry sections provided no continuous barrier. This tended to confirm an earlier study made by the Louisiana Department of Highways [11]. Subsequent investigation further indicated a variation of plasticity indices between the sections, perhaps adding additional dimensions to the analysis.

The Louisiana study examined three ways of adding lime to a subgrade [12]: first, electroosmotically; second, by a pressure injection of a lime slurry; and third, by inserting lime in a posthole. Only the pressure injection method showed signs of success. Tests 2 and 4 years after the injection indicated that the lime had penetrated only from $\frac{1}{2}$ to $1\frac{1}{2}$ inches from the injector tip. The conclusion was that the treatment was not successful and not economically feasible.

In Hudspeth County in far west Texas, a section of IH 10 in 1984 had a rehabilitation contract for 50,098 feet of DVFMB bid at $13.28 per linear foot. When the rural freeway had been constructed 20 years earlier, the activity of the bentonitic clay subgrade was recognized. Those plans had called for 5-foot overexcavation of the clay replaced by a lower PI, less expansive material. The four-lane divided highway had flexible base and an asphaltic concrete pavement. Pavement distortions and regular maintenance had reached a level that rehabilitation with DVFMB was considered necessary.

The contractor used a Vermeer 600 trencher to cut the ditch for the geomembrane, a Du Pont 3358 EVA–coated polypropylene. The average rate of daily placement was 500 feet. Following construction, profilometer readings indicated that the control sections had the smoother riding pavement than the rehabilitated section. Soil borings indicated a possible reason for the pavement results. The bentonitic clay was found to be within 2 feet of the base in several locations. Either the clay had managed to work its way upward, or it had not been replaced as called for in the original construction. In either case, it would provide a good argument for predesign testing to determine actual needs. Twelve psychrometers were also placed on the project to evaluate the rehabilitation. However, they had too short a life span to provide any useful information.

Another IH 10 rehabilitation project going to contract shortly after the Hudspeth County DVFMB work was in the Culberson and Jeff Davis counties (Fig. 10-10). Also in west Texas the rural four-lane divided freeway had suffered considerable asphaltic concrete pavement distortion. In addition to the level-up and final pavement riding course, the rehabilitation called for 40,442 feet of DVFMB. The contractor chose to use a Phillips Petromat MB geomembrane bid at $19.95 per foot in place, complete. The trench for the geomembrane was cut with several backhoes.

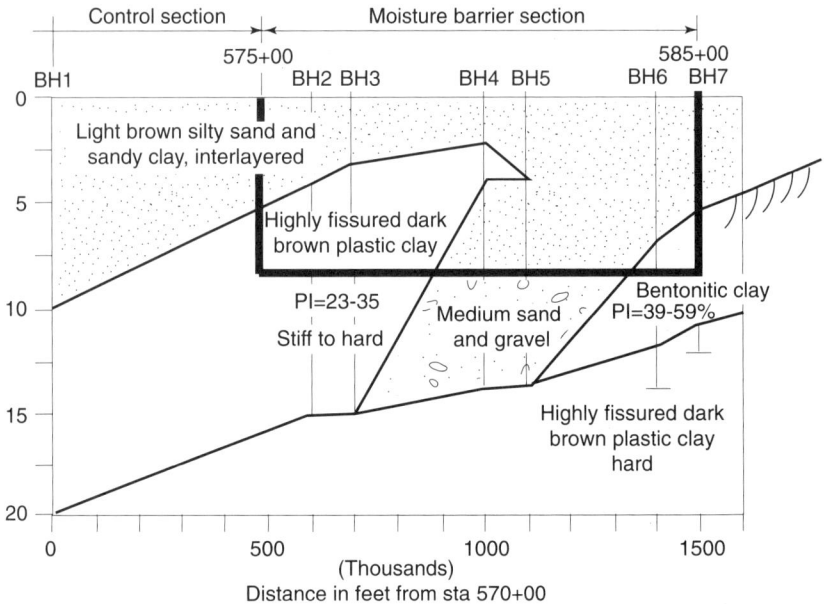

**Figure 10-10.** IH 10, El Paso District, Texas, soil profile.

Average daily placement was 400 feet with a one-day high of 1000 feet. The pavement rides well today with several exceptions. An investigation of the problem areas revealed a discrepancy in the as-built plans. The DVFMB had been left out of several locations to permit vehicle entry and exit from safety rest areas. Those locations were where pavement distortions took place. Another item of concern on this project was reports of the difficulty in removing rock to excavate the DVFMB trench. The lesson learned was that since many decisions need to be made in the field, predesign testing should be thorough [13].

Meanwhile back in south-central Texas, three farm-to-market (FM) highways extending northward from IH 10 between San Antonio and Seguin had rehabilitation contracts. In each case the flexible asphalt surface treatment pavements had been severely distorted by expansive clay subgrades. In addition to an asphalt level-up and finish surface, corrective treatment called for a DVFMB to be placed 8 feet deep along their shoulders. The FM 465 project used 3100 feet of Typar T 063 geomembrane supplied from the State Department of Highways and Public Transportation (SDHPT, now TexDOT) maintenance stocks. For 2 years after its placement in 1985, the pavement remained free of distortions. Since then, several level-ups have been required. The resident engineer has suggested that the geomembrane placed on top of a hill was not placed deeply enough. FM 725 was rehabilitated in 1986, and a Phillips geomembrane was the contractor's choice for a DVFMB. Despite the placement of 6840 feet along the shoulders, the pavement, like FM 465, has also required repeated level-ups, and the resident engineer has suggested that the geomembrane was not placed deeply enough. Later in a 1989 rehabilitation of FM 1516, a Phillips DVFMB was placed 14,400 feet along the pavement's shoulders. The resident engineer's report 4 years later said that the pavement was holding its own [14].

IH 10 between FM 1516 on San Antonio's east side and Seguin is a four-lane divided rural freeway section. Built in the 1960s, the main lanes have a flexible-pavement design. The expansive clay subgrade had received 6 inches of lime stabilization, followed by foundation and flexible-base courses and hot-mix asphaltic concrete pavement. Smooth riding initially, the pavements soon began to crack and suffer distortion from the expansive subgrades. Maintenance costs on a 13-mile section averaged $100,000 to $200,000 annually. In 1985 the section from FM 1516 east to Cibolo Creek had a rehabilitation contract awarded with the goal of reducing the pavement maintenance expenses. The 131,200 linear feet of DVFMB was bid at $13 per linear foot, and it was placed 8 feet deep along both shoulders of the eastbound and westbound main lanes. The geomembrane was to be placed generally through the cut sections where the heaving usually occurred for approximately half the total project length.

The contractor elected to use a Phillips Geoseal geomembrane and a trenching machine for the excavation that was fed by boom attachment to awaiting haul trucks. Daily average placement was 900 linear feet with a one-day high of 2195 feet. The geomembrane trench backfill was specified as coarse limestone screenings. When this source of materials was exhausted, the contractor requested and received permission to use finer limestone scalpings. Shortly after construction, the backfill sank in several locations, leading to shoulder cracking. The cracks were 1 to 6 inches deep, 2 to 6 feet long, and 4 feet into the shoulders. Nondestructive testing revealed voids in the backfill of the finer limestone material, possibly resulting from vehicle-induced vibration consolidation. A remedial contract injected a slurry into the voids, creating a stable condition. A 1993 report indicated little pavement distortion and a significant reduction from the prior maintenance expenses. A further fact to consider on this IH 10 project is the areas where the geomembrane was placed, usually in cut sections, and the areas where it wasn't placed. The challenge of shorting the length of geomembrane placement should be carefully considered. The method chosen to compact backfill material must obviously be carefully considered [15].

The next IH 10 section to receive a DVFMB between San Antonio and Seguin was on the approaches to Santa Clara Creek in Guadalupe County. The pavement was a flexible design, with lime-stabilized subgrade, foundation and flexible-base courses followed by asphaltic concrete pavement. The section began to suffer considerable distortions. The contractor chose a REEMAG, formerly Typar, geomembrane for the DVFMB. The 1987 project used 12,000 feet of the fabric bid at $15 per linear foot. Six years later the resident engineer reported that they no longer had any problem with the pavement.

The first of the next two IH 10 projects extended from Cibolo Creek to Santa Clara Creek, and the second one extended from Santa Clara Creek to the outskirts of Seguin. Both highway sections had required constant maintenance due to heaving pavements caused by expansive soils. The two contracts awarded in 1988 used 14,000 feet of REEMAG bid at $11.42 per linear foot for the first project and 52,800 feet of Phillips geomembrane bid at $11 per linear foot for the second. The work from Cibolo Creek to the approaches of Santa Clara Creek used a graded aggregate trench backfill. The second project toward Seguin also included work on U.S. 90. A trencher was used for the excavation, and a graded aggregate backfill was used. Average daily production was 700 feet. Between 10 and 15 percent of the trench sections suffered caving in of the walls. The trench was moved away from the shoulder, and additional fabric was added to connect the geomembrane in the trench and the shoulder. Five years later reports stated that the pavements were holding up well [15].

SH 97 south of San Antonio, just west of Pleasanton, experienced significant pavement cracking and distortion. The two-lane rural highway with a flexible-pavement design had a rehabilitation contract awarded on it in 1986. A DVFMB placed 8 feet deep was included in the work. The 5600 feet of Typar T 063 was placed along the highway shoulders. Shortly after the rehabilitation was completed, the pavement began cracking again. Further postconstruction testing revealed considerable cracking in nearby facilities. One suggestion has been made that subsurface faulting is to blame. This problem can be prevented with thorough preconstruction testing [16].

U.S. 87 in southeast Bexar County just outside the San Antonio, Texas, city limits is a rural four-lane divided freeway section. It was built over a black clay subgrade that is more suitable for growing cotton than for supporting a highway, and it is adept at distorting pavements. The freeway subgrade was initially treated with 6 inches of lime, two 6-inch flexible-base courses, an asphalt seal coat, and an asphaltic concrete pavement. A 1987 widening and reconstruction contract included 45,000 feet of DVFMB bid at $13.50 per foot. The Typar EVA geomembrane that the contractor chose was placed at a daily rate of 800 feet. No problems were experienced during or after construction [17].

In San Antonio, Texas, two city streets were rehabilitated by TexDOT who placed geomembranes horizontally on the subgrades. The rebuilding of Colorado Street used 2000 feet on the subgrade. The work, which included a flexible asphaltic pavement section, was completed in 1988 and 5 years later was reported to be doing fine. The adjacent section of the street, rebuilt by others, received no geomembrane on its expansive soil subgrade and again suffered pavement distortion. The next street rebuilt by TexDOT was Guadalupe Street. Between 1000 and 2000 feet of geomembrane was placed horizontally on the expansive clay subgrade. It continues to perform well. Both projects used a Mirafi geomembrane [17].

A rehabilitation in the Dallas area of Loop 635 in the MacArthur Interchange included the use of a DVFMB. The geomembrane was placed 8 feet deep in the expansive Eagle Ford shale. The Loop 635 main lanes are in a depressed section in the interchange and have suffered substantial pavement distortion. A 1988 contract called for the placement of 3836 feet of a geomembrane. Initial reports stated that the pavement movements continued and expressed the opinion that the fabric had not been placed deeply enough. Two years later a ramp revision used 700 feet of DVFMB on the interchange area. The resident engineer's assessment was that these applications were not good tests of a geomembrane's ability to control expansive soil movements as the segments were too short [18].

U.S. 84 northwest of Snyder, Texas, received a rehabilitation contract on its four-lane divided rural freeway. Built over an expansive clay subgrade, it developed severe pavement distortions due to the swelling soil, a red clay with PIs in excess of 50. The 1990, $3,406,000 low-bid contract provided an asphalt level-up and finish course for the existing flexible pavement and 46,000 feet of a DVFMB bid at $10.50 per linear foot. The geomembrane was placed 8 feet deep along the shoulders of the northbound and southbound lanes with the usual 2-foot tack of the fabric to the paved shoulders. The geomembrane was placed in three separate segments of 4000, 2000, and 600 feet with areas selected based on where the existing pavement heaves seemed worst. The contractor used a Vermeer trencher with a daily average placement of 1000 to 1500 feet, achieving a one-day high 2500 feet. Variations were made from the specified material, a Grade 3 concrete aggregate with a 1-foot cap of Grade 2 sack-cement-stabilized base was used to backfill the geomembrane's trench. In one section the material excavated from the trench was used for backfill; in another section, blow sand was used. Neither seemed to cause a problem.

These results may further indicate that the backfill selection process is less sensitive than initially considered. The contractor developed a home-grown wheel compactor to tighten the trench backfill (Fig. 10-11). This equipment choice was determined to be unsuccessful. Some caving of the trench walls took place, but that was solved by reducing the delay

**Figure 10-11.** A trench compactor on a Texas highway.

in placing the backfill. Moisture sensors were again used, but they revealed little useful data because of their short life spans. The report after 3 years of contract completion indicated that no pavement problems were showing up. In 1997, 7 years after construction, the resident engineer noted that some pavement heaves had appeared but no maintenance work had been done there yet [19].

The IH 45 widening in the Waxahachie area involved the largest quantity of geomembranes used on a single Texas highway project (Fig. 10-12). This multi-million-dollar improvement began in 1990 on the main freeway connecting Houston and Dallas, and it has used an estimated 285,000 feet of geomembrane. The contractor bid the geomembrane item at $6 per foot for a total cost over $1.7 million. The backfill material initially specified was a coarse sand, but that was later modified at the contractor's request with a price adjustment to $6.25 per foot. The geomembrane used was a Mirafi fabric coated at another company's plant. Early problems with trench walls caving in were resolved by reducing the delay in placing the backfill. Daily placement rates ranged from 1500 to 1800 feet with a one-day high of 2400 feet. Reports were that the geomembrane is working well. The residency elected to use the geomembrane instead of removing and replacing 5 feet of the expansive clay subgrade as originally recommended [20].

A 1996 report on the IH 45 project discusses the setting of four moisture reading locations [21]. A Troxler Sentry 200-AP moisture measuring device was selected. These measuring devices were set in a 2-inch (50-mm) polyvinyl chloride (PVC) access tube, in a hole drilled to the

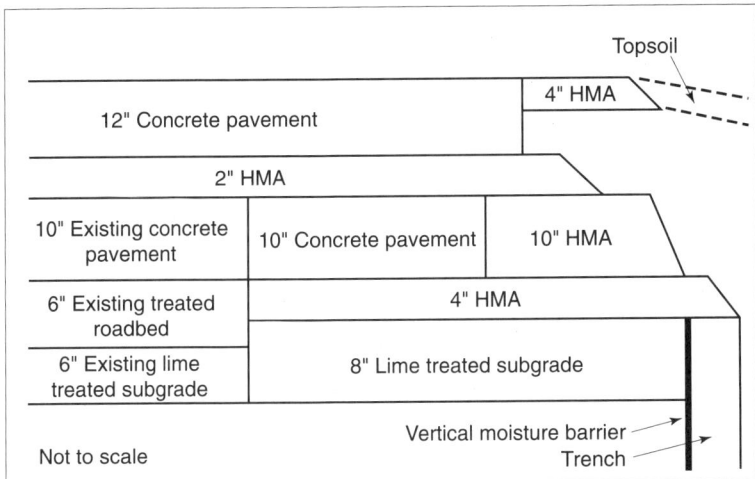

**Figure 10-12.** IH 45 typical section.

desired depth of 10 feet (3.5 m). Readings over a 3-year period were taken at regular intervals. Moisture levels tended to stay consistent within the geomembrane-protected sungrade. Outside the fabric, at three of the locations, the moisture variations were greatest and most pronounced. The fourth location experienced water rising in the PVC pipe, which was thought to come from a previously unidentified water-bearing sand layer.

This 1996 report notes several problems related to the backfill material as well as to the relationship to the earlier pavement section. The density of the backfill is viewed as low due to the difficulty in compaction. The backfill trench was 10 inches (250 mm) wide, and the backfill material was changed several times until a "self-compacted" pea gravel was found. This report observes that the narrowness of the trench, 250 mm (10 inches), made compaction difficult at best. The report goes on to say that the low-density, high-permeability backfill material chosen could make the trench become a water storage area. This could have a negative impact on the efforts to minimize moisture change in the expansive clay subgrade. The report raises the question as to whether this construction technique used with asphalt pavement in a flexible-pavement system could possibly lead to shoulder cracking and sloughing. Generally this was not found to have been the case on similar projects, so perhaps the pea gravel backfill created a unique condition.

The existing pavement on IH 45 did present a challenge. The 10-inch reinforced-concrete pavement rested upon 6 inches of treated subgrade. Beneath was 6 inches of lime-treated subgrade. The widening IH 45 used a 12-inch reinforced-concrete pavement atop 2 inches of hot-mix asphaltic concrete level-up. Below lay a 10-inch concrete roadway with a 10-inch hot-mix asphalt shoulder, both on top of a 4-inch hot-mix asphalt course. All of this work was built on an 8-inch lime-treated subgrade. The DVFMB extended 8 feet down from the top of the lime subgrade treated course. This placed it about 2 feet below the roadway surface. Concerns have been raised about the appearance of water between the edge of the concrete pavement and the asphaltic concrete shoulder, and this is being watched carefully. The same report further suggests that the interface between the concrete asphalt bond may be permitting the infiltration of surface water. At the northern end of the project, water-bearing sands were creating a problem. It was thought that they were allowing water to be trapped inside the barrier, and it was suggested that the barrier be removed in this area. The report does conclude that the moisture barrier on the project appears to be doing what it was designed to do—namely, stabilize the subgrade moisture content underneath the pavement.

In the spring of 1995, the next section of widening and rehabilitation of IH 45 went to contract. It was decided to replace the expansive clay subgrade with a nonexpansive, low PI material. The subgrade removed ranged from 36 to 52 inches based on potential vertical rise calculations. The bid prices for the excavation were $1.92 per cubic yard, and for the select fill embankment, $5.11 per cubic yard. The old reinforced-concrete pavement is being recycled and used as a roadway base [20].

In the early 1960s the Uniroyal Company built a motor vehicle test track north of Laredo, Texas. The track was circular and 5 miles long, and some of it developed problems requiring several rehabilitations and consultations. A 1983 rehabilitation contract called for placing 7800 feet of DVFMB 8 feet deep along the outside shoulder of the track and 13,400 feet of drains along the inside shoulder. A 1985 contract involved asphalt level-ups resurfacing the drains and additional DVFMB. The original recommendation of 9000 feet of geomembrane was reduced by 1200 feet on the contract plans. One segment of the DVFMB was 200 feet long and separated from the remaining geomembrane. A February 1991 study showed that the outside lanes have higher serviceability indices and a smoother ride than the inside lanes. John W. Dougherty, P.E., reported that the geomembrane's DVFMB had eliminated most of the moisture intrusion, which resulted in the smoother riding surface. The track had been sold to Michelin, and no current data are available as to its status. In 1997 Dougherty concluded that the geomembrane might not have accomplished as much good as the underdrain [22].

## 10.3 An Assessment of the Texas Geomembrane Projects

An assessment of the Texas experiences and the testing usually following the geomembrane placements is appropriate (Table 10-7). It represents efforts to control expansive soils with the geomembranes' having been in place 21 years, from 1976. Most of the Texas highway projects involved the use of deep vertical fabric moisture barriers. Since 1976, their rate of placement has increased to one-half mile, 2600 feet, a day, the bid price for the 8-foot depth has decreased from the $20 per square

**Table 10-7.** Geomembranes on Texas Highway Projects

| Highway | Location | Quantity | Bid | Year | Performance |
|---|---|---|---|---|---|
| Gen. McM. | San Antonio | 600 | — | 1976 | Good |
| IH 410 | San Antonio | 4,978 | 20.00 | 1978 | Good |
| IH 37 | San Antonio | 21,584 | 21.00 | 1980 | Good |
| IH 10 | San Antonio | 24,745 | 15.00 | 1983 | Good |
| U.S. 281 | San Antonio | 4,705 | 3.00 | 1983 | Good |
| IH 30 | Greenville | 2,000 | 10+ | 1984 | Questionable |
| IH 10 | Hudspeth County | 50,098 | 13.28 | 1984 | Poor |
| IH 10 | JeffDCulb County | 40,442 | 19.95 | 1985 | Good |
| IH 10 | Bexar County | 131,200 | 13.00 | 1985 | Good |
| SH 37 | Charlotte | 5,500 | 10.00 | 1986 | Poor |
| FM 465 | Guadalupe County | 3,100 | 10* | 1986 | Poor |
| FM 725 | Guadalupe County | 6,840 | 20.00 | 1986 | Poor |
| Cldo. St. | San Antonio | 1,000 | 7.00 | 1987 | Good |
| IH 10 | Guadalupe County | 12,000 | 15.00 | 1987 | Good |
| U.S. 87 | San Antonio | 47,000 | 13.50 | 1987 | Good |
| IH 10 | Guadalupe County | 14,000 | 11.42 | 1988 | Good |
| III 10/90 | Guadalupe County | 52,800 | 11.00 | 1988 | Good |
| FM 1516 | Bexar County | 14,400 | 10.00 | 1989 | Good |
| IH 635 | Irving | 3,638 | 15.80 | 1989 | ? |
| U.S. 84 | Snyder | 46,000 | 10.50 | 1990 | Good |
| Guad. St. | San Antonio | 2,000 | ? | 1990 | Good |
| IH 45 | Waxahachie | 285,140 | 6.00 | 1990 | n.e.y. |
| IH 635 | Irving | 2,000 | ? | 1991 | ? |

*First.

yard or running foot to the $6 range. The results of using geomembranes have been overwhelmingly positive. Maintenance costs, when reported, are down, and the engineer's image, as well as the contractor's and the design team's image, is up.

Postconstruction studies, most frequently conducted in Texas by the Texas Transportation Institute, have monitored several of these projects. They have provided additional information on the nature of the cracking of the clay and the root depth as a guide for placement depth. They also address the need to consider the nature of the area as to whether it is arid or semiarid before deciding whether to use a deep vertical fabric moisture barrier. Generally, they have concluded that the DVFMB does a good job and fulfills the purpose for which it has been designed. Most of the engineers involved in the projects in Texas would agree with that assessment.

## 10.4  References

1. Fu Hua Chen. *Foundations on Expansive Soils.* Elsevier, Amsterdam, 1975.
2. E. B. McDonald. *Experimental Moisture Barrier & Waterproof Surface.* South Dakota Department of Highways, Pierre, October 1973.
3. H. C. Porter. "Observations of the Texas State Highway Department on the Subsequent Effects of the Uniformity and the Nonuniformity of Foundation Soil Types of Pavements; and Also the Effects of Uniformity and of Nonuniformity of Moisture Content Fluctuations in Soil Foundations of High Volumetric Change." *Proceedings of the International Conference of Soil Mechanics and Foundation Engineers,* Harvard University, Cambridge, 1936.
4. H. C. Porter. "Roadway and Soil Mechanics Data—Part XI, Moistures in Clay Soils Beneath Pavements." *Bulletin of the Agricultural and Mechanical College of Texas,* College Station, 1942.
5. B. A. Brakey, "Road Swells, Causes and Cures," *Civil Engineering,* American Society of Civil Engineers, New York, 1970.
6. C. M. Kelley, "A Long Range Durability Study of Lime Stabilized Bases." National Lime Association, Arlington, Va., 1988.
7. T. C. Feng and M. B. Clisby, "Experimental Highway Construction Techniques for Active Clays in Mississippi," Mississippi Highway Department, Jackson, 1974.
8. R. L. McKinney, J. E. Kelly, and C. McDowell. *The Waco Ponding Project.* Research Report 118-7. Center for Highway Research, University of Texas, Austin, 1974.
9. M. L. Steinberg. *Ponding on Expansive Clay Cut.* Transportation Research Record No. 661. Transportation Research Board, Washington, D.C., 1977, pp. 61–66.

10. M. L. Steinberg. *Departmental Research Report No. 187-12.* Texas Department of Highways and Public Transportation, Austin, 1985.
11. M. L. Steinberg. *Deep Vertical Fabric Moisture Barriers under Swelling Soils.* Transportation Research Record 790, Washington, D.C., 1981, pp. 87–94.
12. M. L. Steinberg. *Monitoring the Use of Impervious Fabrics, Geomembranes, in the Control of Expansive Soils.* Departmental Research, Report No. 187-12, SDHPT, 1985.
13. R. E. Magers. Unpublished data, 1992.
14. D. A. Gay and R. L. Lytton. "Development of a Predictive Model for Pavement Roughness on Expansive Clay," a doctoral dissertation, Texas A&M University, College Station, 1991.
15. C. M. Higgins. *Lime Treatment at Depth.* Research Report No. 41. Louisiana Department of Highways, Baton Rouge, 1969.
16. M. L. Steinberg. *Further Monitoring of Twelve Geomembrane Sites in Texas.* Departmental Information Exchange, State Department of Highways and Public Transportation, Austin, 1989.
17. D. R. Stein. Unpublished data, 1992.
18. D. R. Stein. Unpublished data, 1993.
19. G. Clement. Unpublished data, 1992.
20. R. E. Magers. Unpublished data, 1993.
21. C. Tucker. Unpublished data, 1982.
22. M. Taylor. Unpublished data, 1997.
23. H. Stanford. Unpublished data, 1997.
24. H. S. Bredenkamp, T. T. Scullion, and T. J. Ragsdale. *Evaluation of the Vertical Moisture Barrier.* Texas Transportation Institute, Texas A&M University, College Station, 1996.
25. J. W. Dougherty, P.E. Unpublished data, 1997.

# 11
# Case Studies: Highways, Roads, and Streets beyond Texas

Though Texas was one of the earlier states to use geomembranes in efforts to control expansive soils in construction, there were others involved in similar projects. Wyoming, in fact, has twice the number of geomembrane projects on its highway system as Texas. Many other states have used geomembranes, usually as deep vertical fabric moisture barriers, many in combination with horizontal barriers. Over 100 projects have been identified around the world. Usually the report is that geomembranes have been successful in controlling expansive soil.

## 11.1  Wyoming

The Wyoming Department of Transportation (WyDOT) began using geomembranes in 1981 in the first of more than 50 projects in the state (Fig. 11-1). Until this time, they had been blasting the expansive Fort Union shale and recompacting it to cope with highway damages, but they concluded that the results were less than satisfactory. On an early

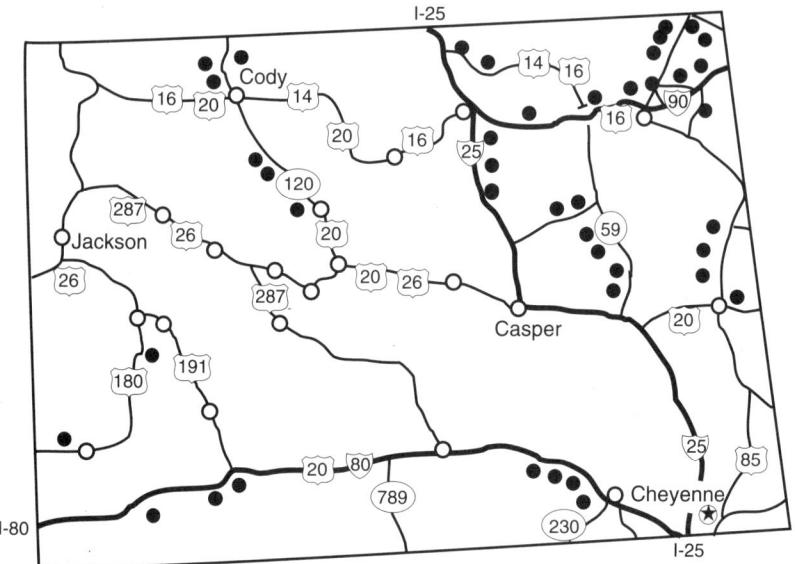

**Figure 11-1.** Wyoming geomembrane locations.

project near Kaycee, they had tried blasting and recompacting the Fort Union tertiary shale, but the results were disappointing. The Fort Union shale is a younger, fairly weathered shale in contrast to the Cody, Pierre, and Steele shales that are also abundant in the state.

On a rehabilitation project on IH 80 between Laramie and Wolcott, 106,854 square yards of a coated Typar geomembrane was placed horizontally on a 1-mile section of the existing 30-year-old pavement that was undercut 5 feet. The geomembrane was placed at the bottom of the undercut segment and covered with 5 feet of saturated backfill, which was followed by base and pavement. Twelve years later a 1993 project assessment indicated that the geomembrane kept the water out of the shale, reducing heave and frost damage (Fig. 11-2). In other projects

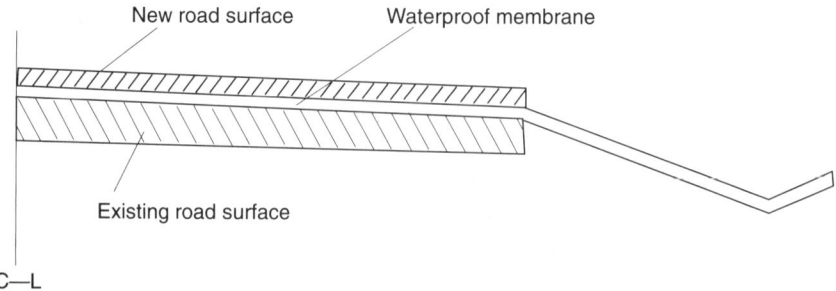

**Figure 11-2.** First Wyoming geomembrane section.

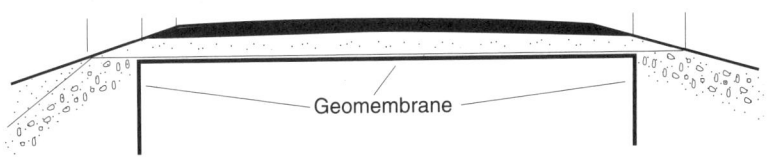

**Figure 11-3.** Wyoming vertical and horizontal geomembrane section.

between 1982 and 1986, an estimated 227,950 square yards of geomembranes were reported to have been placed on Wyoming highway projects [1].

By 1986 WyDOT was using geomembranes horizontally and vertically on their rehabilitation projects (Fig. 11-3). IH 90, SH 59, and SH 112 received geomembranes in 1987. One project used Typar 3353, another used Typar 3358, and the third used MCF 1212. On all three, the geomembrane was placed horizontally over the existing pavement, sometimes followed by 6 inches (150 mm) of sand, then base and pavement. Vertical placement ranged from 2.5 feet to 4 feet (0.7 to 1.2 m) deep.

In 1988 the department awarded 10 contracts that called for geomembranes. Project geomembrane totals varied from 160 to 77,811 square yards. Eight projects used MCF 1212, one used TJDD FABRENE, and the other, WI9. The next year geomembranes were used on three rehabilitation projects. On IH 25 between Kaycee and Buffalo, 29,822 square yards of WI 9 geomembranes were used horizontally and vertically. Two other projects in 1989 used 55.070 square yards (46,039 m$^2$). One used MCF 1212 while the other was not identified. All used the horizontal and vertical placement methods.

Ten additional 1990 WyDOT rehabilitation contracts also called for placing the geomembrane both horizontally and vertically. One of the larger projects, SH 315 from Mountain Street Lane to Decker Road, used 101,774 square yards. Two other projects, U.S. 212 and U.S. 85, used 233,000 and 210,210 square yards, respectively. In 1991 Wyoming contracted five projects using geomembranes. Quantities on the contracts varied from 200 to 218,300 square yards. The U.S. 85 contract between Mule Creek Junction and Newcastle used 195,880 square yards of geomembranes.

The 1992 WyDOT rehabilitation contracts involved seven projects using geomembranes (Fig. 11-4). SH 59 between Douglas and Gillette included 230,000 square yards of geomembranes. All the other projects were on roadway rehabilitation areas except one where the geomembrane was used to line a ditch. The resident engineer at Newcastle reporting on the 1992 contract for rebuilding the Pine Haven county road to state standards, SH 113, stated that the geomembranes were

**Figure 11-4.** A moisture barrier Wyoming project.

placed horizontally and vertically. The vertical placement was 3 feet deep along both shoulders. No moisture determinations were made. Excavation for the DVFMB was accomplished using Vermeer and Ditch Witch trenchers. The trenches were 6 to 8 inches wide, and the backfill used the excavated material. Backfill compaction was usually accomplished with homemade compactors (Figs. 11-5 and 11-6). Daily production of the geomembrane placement depended on the base being hauled to cover the horizontally placed geofabric; placement thus aver-

**Figure 11-5.** Compacting the trench backfill on a Wyoming highway.

**Figure 11-6.** Trenching for a vertical barrier on a Wyoming highway.

aged about 3000 feet. Daily production was limited by the need to cover the horizontal geomembrane and avoid leaving it uncovered overnight. The 1997 reports indicate that the SH 113 pavement probably needs additional maintenance [2].

In 1993, five geomembrane projects went to contract. Fabric quantities varied from 109,000 to 254,000 square yards. In 1994 and 1995, 10 more projects calling for geomembranes were awarded. Quantities of the geomembranes used were 324,000 square yards in 1994 and 221,500 square yards in 1995. In 1996, 8 contracts were awarded using 315,582 square yards of geomembrane, placed horizontally and vertically. Between 1987 and 1996, a total of 4,188,142 square yards of geomembranes had been used on WyDOT projects. Average annual prices varied from $2.33 to $0.72 per square yard. Placing the geomembrane 3 and 4 feet deep seemed to work fine, but it was noted that the need to go deeper might exist. Rainfall in the state varied from 8 to 10 inches annually. The trench backfill is usually the material excavated, and compaction is achieved by using a wheel compactor. A 1997 retabulation, Table 11-1, indicates that there were 54 geomembrane WyDOT projects initiated between 1986 and 1996 [3]. Table 11-1 is a 1997 recapitulation of the geomembrane placements on WyDOT highways. The projects between 1981 and 1986 have not been included here although they were noted elsewhere in the WyDOT review.

The WyDOT resident engineer at Sundance, Wyoming, in a 1997 conversation, described the northeast area of the state as "bentonitic min-

**Table 11-1.** WyDOT Geomembrane Projects from 1986 through 1996

| Year | Quantity, square yards | Projects |
|---|---|---|
| 1987 | 58,120 | 2 |
| 1988 | 9,900 | 1 |
| 1989 | 191,000 | 3 |
| 1990 | 622,100 | 6 |
| 1991 | 418,100 | 4 |
| 1992 | 1,135,990 | 10 |
| 1993 | 881,000 | 10 |
| 1994 | 324,000 | 5 |
| 1995 | 221,150 | 5 |
| 1996 | 318,562 | 8 |
| Total | 4,188,142 | 54 |

SOURCE: M. B. Hager, unpublished data, 1997.

ing country." The heavy soils are classified as A-7s with PIs often in the 90s. The geomembranes' vertical placement depth is usually 4 feet, or below the ditch bottom. On U.S. 212, which crosses the state for 22 miles between Montana and South Dakota, WyDOT placed a geomembrane to keep the subgrade moisture constant. The work on U.S. 212 was awarded in three separate contracts. Adjoining sections of the highway not receiving the geomembrances were later described as being in poor shape. WyDOT used geomembranes in all their projects in the northeast part of the state. The results have generally been very satisfactory [4].

## 11.2 Other States

The Arizona Department of Transportation (AZDOT) has used geomembranes on a variety of projects over a period of years. The first in 1992 was on IH 40 between Needle Mountain Road and SH 95. The horizontal placement extended from subgrade crown to ditch backslope. The contractor's bid price was $10 per linear foot for the estimated 4400 feet of geomembrane. Previous roadway distortions have not reappeared. SR 347 had 3 feet overexcavation of the subgrade with horizontal placement of the geomembrane. Despite the highly plastic clays, the pavement has not been disturbed over the 1-mile geomembrane section. Three other AZDOT projects used geomembranes horizontally and vertically. IH 40 between the Apache County line and the

Piute Road was completed in 1995. Bid price for the geomembrane, placement, and backfilling was $10 per foot. SH 188 between Ash and Sycamore Creeks had a similar rehabilitation contract. A second contract on the same road extended from Ash Creek. Three other highways are identified in the 1995 report as being planned and calling for the use of geomembranes. They are IH 40 east of Holbrook, SR 95 in the Mojave City area, and SR 666 in the Lutresso area.

The U.S. Bureau of Indian Affairs (BIA) used geomembranes on reservation roads in the Arizona and New Mexico area. Four roads were on the Navajo reservation where the geomembrane was placed horizontally. Two projects used Mirafi MCF 1212, and two used Typar 3353. Another BIA office reported having used geomembranes; one of its projects had an 8-year history, and the geomembrane has continued to serve well. The BIA is generally pleased with the performance of geomembranes in reducing the destructive movement of the expansive soils, and they are continuing with their use. They usually place the geomembrane horizontally from ditch backslope to backslope [5].

The Montana Highway Department has used geomembranes in the rehabilitation of three projects. For IH 90, in a section adjoining the Wyoming state line, the geomembrane was placed 3 to 4 feet (0.9 to 1.2 m) deep along the shoulders of the four-lane divided rural freeway. This highway section is in the Big Horn uplift area, and the subgrade was undercut 2 to 3 feet (0.6 to 0.9 m). The geomembranes were placed on several sections of the 23-mile (34.5-km) contract, which was completed in 1986. Seven years later the pavement was reported to be in good shape.

A rehabilitation contract on Montana's IH 15 in the Great Falls area had 5 feet of subgrade excavated and redensified, and the geomembrane was placed vertically and "probably horizontally." A Vermeer trencher was used on the project, and it experienced trouble in excavating below 4 feet (1.2 m). Another IH 15 project north of Great Falls included overexcavating 2 feet (0.6 m) deep, and replacing the material with sand. A geomembrane was placed vertically along the shoulders of the four-lane freeway. Little postconstruction followup was reported along with the comment, "There seems to be little interest in dealing with the expansive soils."

Two California Department of Transportation (CALTRANS) projects in the Sacramento area called for the use of geomembranes in its rehabilitation contracts. The interchange between IH 80 and U.S. 50 was built in 1970. Shortly after construction completion, pavement and slope distress became apparent. Studies began in 1981, which culminated in a rehabilitation contract in 1987. The work included removing the distressed pavement, placing a geomembrane 2500 feet horizontally and 6 feet deep vertically. The trench was backfilled with 4 feet of field sand and 2 feet of clay. A new pavement structural design followed, which included a dense asphaltic concrete pavement. Four years later the pavement was doing fine, but the dike built to control drainage from the slope wasn't performing satisfactorily. Another contract addressed this problem, and it called for some additional geomembrane to be placed horizontally. The CALTRANS report indicated that there is a need to take care of geomembranes during construction operations. It was also noted that it would be instructive and beneficial to secure funding to conduct postconstruction testing and observations.

The New Mexico DOT has used the geomembranes on two interstate highway projects. IH 40 west of Albuquerque in the vicinity of Rio Puerco had 2 feet of subexcavation, and the Phillips MB geomembrane was placed horizontally over the subgrade and turned down vertically 2 feet. The trench was backfilled with native material [6]. A rehabilitation contract in 1996 on IH 25, near Hatch, MPs 37–40, used a 4-foot deep, vertically placed geomembrane. Both projects are doing well. A proposal to use geomembranes on a section of IH 25 north of Las Cruces was deferred.

The Mississippi DOT is another state agency using geomembranes horizontally in efforts to control expansive soils. An IH 20 rehabilitation contract in Scott County used the fabric on 13 locations of the eastbound lane and 14 locations on the westbound lane where the pavements had shown the most heaving. A Phillips Petromat MB II was placed 9 feet from the roadway centerline in each direction to 6 feet beyond the ditchlines. A total of 142,820 square yards of geomembranes were used, bid at $2.33 per square yards for a sum of $324,198. Construction was completed in 1990. The assessment in 1997 shows that there has been little difference in the moisture contents of the membrane and the control sections subgrade. A full report is being prepared and will be released shortly [7].

The Utah DOT has used a geomembrane, a Typar 3401, on a 5-mile rehabilitation of U 10. The geomembrane was placed horizontally across the highly expansive Mancos shale subgrade. The estimated 114,222 square yards were bid at $0.90 per square yard. The project was originally built in 1972, and through the years the pavement has suffered considerable distortion and distress. The rehabilitation contract in 1981 has resulted in a pavement with no swelling or heaving, a cost saving in maintenance expenses, and the elimination of public criticism [8].

The Nevada DOT (NDOT) has used a geomembrane horizontally on a U.S. 385 rehabilitation project near Reno. Prior to this rehabilitation effort, they had been removing and replacing the expansive soils with nonexpansive borrowed material on several projects. Their assessment, however, of this remove-and-replace approach was that the swelling and heaving was beginning anew after a while. On the U.S. 395, 3000 to 4000 feet had been specified for a remove-and-replace project in the contract. In rehabilitation projects elsewhere, the geomembrane sections were said to be doing well 3 years after project completion [9]. A 1997 update indicates that NDOT plans to use geomembranes on three upcoming projects. The geotextile will be placed vertically 5 feet deep along the roadway shoulders. One proposed project is the IH 80 rehabilitation in the Elko area. Another planned rehabilitation contract where geomembranes will be used is U.S. 50 near Dayton. New construction on the SH 395 interchange in Reno is planned where the geomembrane will be used vertically. In all three cases native material will be used for the trench backfill [10].

Kentucky's western sections have some of Porter's Creek expansive clays. The land is fairly flat, and "people get very upset about bumps and heaves in the roads." IH 25 in the Paducah area was rehabilitated in 1990 using geomembranes after removing the pavement and undercutting the expansive subgrade 5 feet. The clay had a liquid limit of about 100 with 50 to 70 percent moisture in its natural state. The Mirafi geomembrane was placed 8 feet deep and then lapped 5 feet along the shoulders. Most of the geomembrane sections were placed along the roadway in maximum lengths of 500 feet. The project is considered to be doing well [11].

A 1997 report from the South Dakota Department of Transportation reviewed the work done on a 3-mile section of U.S. 212. The test included eight sections, one serving as the control, two with undercutting the subgrade 3 and 5 feet, followed by reconstruction, and one using geomembranes $3\frac{1}{2}$ feet deep along the highway shoulders. Geomembrane sections were also placed horizontally across the old roadbed. The trench was cut with a Ditch Witch machine, a heavier model than the one used initially due to the difficulty in cutting the Pierre shale. Other tests in various sections included grinding the existing pavement, base and recompacting it, then repaving the roadway, which is now viewed as the most cost-effective solution.

A review of the construction history for U.S. 212 in South Dakota showed that it was built during a period of prolonged drought. Later, when the soil absorbed a normal to high amount of moisture, the pavement heaved 3 to 5 inches in some places. Plans to repair the test section were developed in 1989 and 1990, and construction followed in 1992. Current assessment indicates that the undercut reconstructed section behaved as well as the geomembrane section. The geomembrane was supplied by Midwest Canvas to the state specification, and it was delivered in rolls 22 by 200 feet. A note on the construction history of this area has been written by South Dakota materials engineer Eugene McDonald, who reports that perhaps 20 years ago a geomembrane was placed in this highway section 4 feet deep along a pavement shoulder. He commented that if the geomembrane had been placed twice as deep, it would have stopped the heaving and swelling. A full report is expected shortly [12].

## 11.3 Australia

Approximately 20 percent of Australia's surface is covered with expansive soils. These soils appear generally along the southern and eastern coastal areas where most of its 18 million inhabitants live (Fig. 11-7). The expansive soils have left visible evidence of their presence in swelling, heaves, and shrinkage of their highways. Australia first used a vertical geomembrane moisture barrier in 1985 as part of a rehabilitation of the Sunraysia Highway in the Morton Plains area Victoria. The test site is 12 km (9 miles) south of Birchip in the semiarid area of the Wimmers region in northwestern Victoria. The aim of the test was to determine

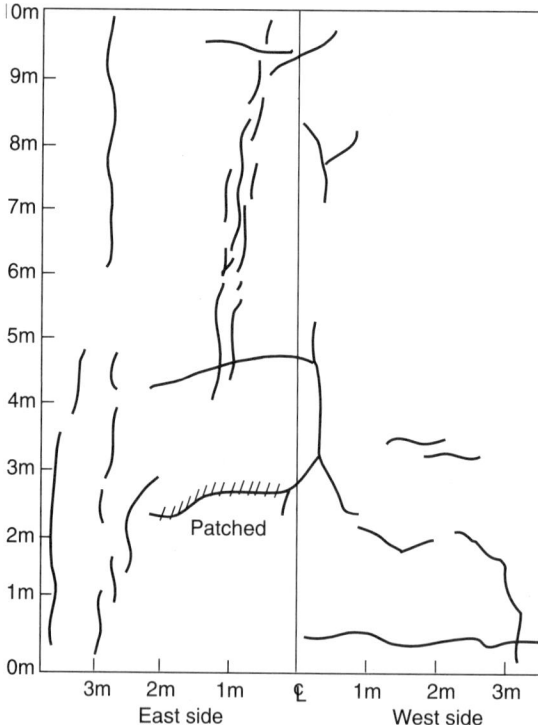

**Figure 11-7.** Australian highway crack survey.

whether the vertical moisture barrier would reduce the pavement distortion caused by the expansive soils.

The highway subgrade had PIs ranging from 57 to 67. Ruts in the pavement were reported to be 40 mm (1¾ inches) deep with severe longitudinal cracking in the highway pavement. Between April and June 1985, the reconstructed highway experienced severe pavement distortion. The section had been widened from 5 to 7 m (16 to 23 feet) with asphalt sprayed to a depth of over 170 to 300 mm (6 to 12 inches) to a weathered sandstone base and subbase. The loss of pavement shape requiring remediation resulted after only 10 years of roadway life rather than the expected 20-year span [13].

Geologically the Morton Plains area is described as similar to the Horsham area in Victoria, Australia. The soil profile includes 0.6 to 2.2 m (2 to 9 feet) of grey to black clay overlaying a heavy white clay. Holden notes that the buried gilgai in the area have high swell potential and decreasing permeability as depth increases. Crediting the work of Kassiff and Holland in 1966, Holden further observes that the test site is low lying, poorly drained, and has experienced flooding several times.

Soil studies were conducted on both sides of the highway. The soil was identified as a highly expansive sandy clay with an LL of 82 to 85, a PL of 19 to 25, and PIs from 57 to 67. The subbase and base of sandstone is very good on placement, but it tends to weather and degrade rapidly when wet. The experimental section of 284 m (925 feet) has trees on both sides of the road.

The reconstruction of Sunraysia Highway took place from April to June 1985. The experimental section had a 130-m (425-foot) control section and a 154-m (500-foot) test section. Trenches were excavated along both sides of the road, 8 m (26 feet) from the centerline (Fig. 11-8). The trench was extended into the control section to cut tree roots, so as to remove that variable from the experimental evaluation. A backhoe was used for the trench excavation cutting 1.5 m (4.9 feet) below the subbase. The 0.4-m (17-inch) wide trench was backfilled with a native clay (Figs. 11-9 through 11-11). A single layer of 0.2-mm-thick black polythene membrane was used (Fig. 11-12). The membrane came in rolls 50 m (325 feet) long and 4 m (13 feet) wide. Two sheets were joined longitudinally to create a 7-m (323-foot) sheet, with the lap of the two sheets. The native clay trench backfill was shoveled in with 100-mm (4-inch) lifts and compacted with water application (Fig. 11-13). The last 300 mm (12 inches) was compacted with a heavy roller. Level readings were taken at the ends of the wet and dry seasons along the road's east side. They indicated that between 1985 and 1988 there was a slightly reduced heaving pattern in the geomembrane-protected section but no significant difference in the control section. A decision was made to place the geomembrane deeper on the next test section.

**Figure 11-8.** Australian trenching for geomembrane placement.

**Figure 11-9.** An Australian trencher.

The second field trial included increasing the depth of the geomembrane placed in two sections of the initial trial. The increase was usually about 1 m (3.25 feet). It also included a third trial section with the vertical barrier placed 2.5 m (8.1 feet) deep. The two sections where they were added to the first placements used an adhesive tape to secure their bonding. Backfilling with the native material excavated from the trench was placed in layers 200 mm (8 inches) deep until the last 400 mm (16 inches), where the layers were reduced to 100 mm (4 inches). A vibratory roller completed the operation.

**Figure 11-10.** The Australian trenching train.

**Figure 11-11.** An Australian trencher.

During the excavation for the test sections, it was observed that the roots seemed to be within 400 mm (16 inches) of the surface. The roots were small, from 1 to 5 mm thick, though some reached 10 mm ($3/8$ inch). Exposing the geomembrane placed in the first trial section indicated that the roots did not penetrate the fabric but, instead, turned horizontally and grew along the barrier. Where penetration did take place, the geomembrane had been torn previously by rock or the placement operation.

Heave measurements followed in 1988 and 1989. The results indicated that the shallower geomembrane installation was not deep enough. The

**Figure 11-12.** An Australian geomembrane carrier.

**Figure 11-13.** Placing a flowable backfill in Australia.

placement 2.5 m (8.1 feet) deep greatly reduced and virtually eliminated roadway heave. Due to the roots' puncturing the 0.2-mm polythene fabric in some places, it was recommended that a thicker geomembrane as well as a root inhibitor be used. Following the Sunraysia project, a vertical moisture barrier was placed on the Princes Highway, located in Little Rock, Victoria. The results of this work have not yet been published.

In a presentation at the UTS Geomechanics Seminar on Road Engineering in Sydney, Australia, the results of the Sunraysia highway test sections were reviewed [14]. The researchers noted the work of Jayatilaka from Texas A&M University who reported that there was little benefit from vertical barriers in tight clays that have a low permeability [15]. Evans and Holden have stated that this was not the case on the Sunraysia Highway where great value was found in the use of the vertical moisture barriers. They did observe, however, that the value of these barriers diminishes if the clays have deep cracks or if water-bearing sand seams are acting as conduits. Suction tests indicated the depth and impact of the tree roots, but major moisture changes were observed to be slightly less than 2.8 m (9 feet) below the surface.

Precise leveling at the Sunraysia test sites indicated that the deep vertical moisture barrier resulted in a significantly reduced heave and a smoother road surface. Additional tests in 1993 and 1994 using computer-reduced data from the Australian Road Research Board's (ARRB) laser profilometer reinforced these conclusions about the benefits of the barriers. Use of vertical barriers resulted in substantially less heave and shrinkage compared to the control sections. A crack survey indicated more roadway openings on the control sections compared to those with the barriers. The geomembrane has reduced the roadway cracking and its damages to the roadway pavement.

In 1991 a test section on Melrose Drive in Tullamarine used a high-density polyethelene (HDPE) geomembrane installed 2 m (6.5 feet) deep and 1.2 m (3.9 feet) back from the median curb. A root inhibitor was placed at the bottom of the 300-mm (12-inch) wide trench. A polyethylene foam was used to seal the backfill around service lines crossing the trench. Three years later a significant decrease in pavement movements where the barriers were placed was reported. It was also noted that the foam is a very expensive item.

The report tells of the beginning of a geomembrane vertical moisture barrier project on the Henry highway, located near Doonen, a small settlement 9 km (6 miles) from Horsham. Two experimental sites have been chosen, one a 1000 m (3250 feet) and the other 600 m (1359 feet) long. Profilometer readings have been taken on the northern test section. Plans include placing a geomembrane 2 m (6.5 feet) deep with a double thickness to 0.4 mm. The geomembrane will be placed using a mobile dispenser. Trench excavation will involve the use of the slim-line model to cut a trench 2 m (6.5 feet) deep and 78 to 94 mm (3 to $3^{3}/_{4}$ inches) wide. Flowable backfill will be used consisting of cement, fly ash, local sand, and water. The report notes that in earlier similar projects, the deep vertical moisture barrier has resulted in less heave and pavement distortion.

A 1997 report provided information on the construction and results of the first of the Henry highway, near Doonen, experimental sections. The initial section was excavated in February 1994. At first an existing conventional trencher was used. The focus was to determine the effectiveness of the geomembrane's mobile dispenser. The second test section was begun in December 1994 using a Ditch Witch 6519 and the slim-line excavator chain. Two 100-m (325-foot) sections were scheduled for barrier placement to be separated from the first by a 100-m control section. The slim-line trencher used on this section was the flowable backfill. The work was to be done along the east side of the pavement adjacent to a tree plantation. Problems arose that were resolved by the redesign of the trencher's teeth, which were then coated with tungsten, and the addition of a crumber bar and a soil chute. A lime dispenser was added to reduce the caking of the wet clay on the trencher teeth.

Evans reports that after adding chisel teeth to the chain, the project work was begun again. However, as work progressed, instead of running into the smaller weathered sandstone cobbles they had first encountered, they began to run into boulders up to 250 mm (10 inches) that blocked further trenching. The boulders lodged in the soil ramp and blocked and derailed the chain. The chisel teeth were bent, snapped, and sheared. The soil ramp was similarly damaged and rendered inoperable.

A third trial in February 1995 completed the placement of the 100-m barrier on the east side of the road. A shorter, stocky rock boom was used to excavate to the bottom of the road base. The slim-line trencher then followed to cut a deeper narrow trench. Evans reports that once the boulders and cobbles were removed, the slim-line trencher operated just fine.

Satisfaction with the Doonen tests led to a 2-km (1.3-mile) vertical moisture barrier section of the Western Highway in the Merwyn Swamp area 12 km (8 miles) east of Kaniva in northwest Victoria. Several modifications were made to the equipment to increase productivity and reduce the use of supplemental equipment if rock were encountered. An antiflex back chain was used to reduce its unwanted pivotal movement, rock teeth were alternated with short tungsten carbide chisel teeth to effectively handle difficult trenching conditions, and an idler sprocket was added to reduce lateral chain movement. A more pronounced chain tongue was installed to eliminate chain derailment. These modifications helped to maintain a trenching and geomembrane placing rate of 60 m (175 feet) an hour. In a current project underway on the Tullamarine Highway near Melbourne International Airport, the use of lime fly ash grout curtains is being studied. The experiment is reported in a master's thesis submitted by Sandra DeMarco at Swinburne University.

The first geomembrane highway project in the state of South Australia was located on Biscay Road in 1984. The geomembrane was placed horizontally and then 1.5 m (4.9 feet) deep vertically. A trencher was used to excavate for the plastic barrier on a 300-m (975-foot) test section. The depth of placement was limited by the trencher's ability to excavate only 1.5 m [16].

Londale Road, a four-lane highway 8 km (5.3 miles) south of Adelaide in South Australia, had suffered extreme pavement distortion and cracking. A 2-km (1.3-mile) section was rehabilitated in June 1995. A deep vertical moisture barrier was used on three segments of this work. Roadside grasses and mature trees along the roadway were viewed as having exacerbated subgrade drying and cracking of the pavement. The geology of the area consists of highly expansive clays over a weathered limestone. A vertical moisture barrier was installed along the sealed shoulder in two locations and along a deep concrete gutter in the third. The slim-line trenching boom cut a 78-mm (3-inch) wide trench using the hard-wearing, tungsten-coated chisel teeth. Powdered lime was applied to the teeth to minimize caking caused by the wet clay. Trench walls collapsed in several areas. Rapid installation of the geomembrane after trench excavation followed by prompt application of the flowable backfill solved the problem. The report notes that no trenches should be left open overnight [17]. The engineer for South Australia's road department views this 1997 project as unsuccessful.

The development of the slim-line trencher and the flowable backfill represents a considerable step forward in the economic feasibility of controlling moisture change and its related volumetric variations in expansive soils. The close cooperation between VicRoads, the ARRB, and the Swinburne University of Technology bodes well for further progress in reducing the damages from expansive soils. Additionally, the two councils in several instances have sought and received aid from VICRoads and particularly Dr. James Holden, and the results of this cooperation have been tremendous.

## 11.4  Israel

Israel has also been using geomembranes on test sections of roads distorted by expansive soils. One project is Road 65, which was rehabilitated in 1989 using a 400-m (1300-foot) geomembrane section (Fig. 11-14). Near Kfar Pazar, a backhoe was used for the geomembrane trench excavation. The roadway, a two-lane highway, is not doing too well. Traffic has grown tremendously on the road, and it is believed that the pavement is suffering because it was not designed for the traffic it carries.

Road 69 in the Beit Alpha area was rehabilitated using a vertical moisture barrier in 1990. The trench excavation was carried out with a backhoe, and two 500-m (1600-foot) sections received the bitumin geomembrane (Fig. 11-15). The thickness of the geomembrane has been indicated to be 40 mils. The pavement today is in excellent shape and presents no

**Figure 11-14.** An Israeli geomembrane project.

**Figure 11-15.** A trench compactor on an Israeli highway.

problem. Additional geomembrane sections have been placed in the Tel Aviv area and on the highway between Tel Aviv and Jerusalem [18].

## 11.5  An Evaluation

There are many positive indications that geomembranes have substantially reduced damages caused by expansive soils, not only in the United States but also in Australia and Israel. Analyses have revealed that when they do fail, it is because they have not been placed to the depth of surrounding roots or they have been placed in dense, tight clay with high permeability. Some fears have been expressed that water will build up underneath the pavement and cause failures, that the geomembrane will not be sufficiently durable, and that this is a new, untested method. None of these fears have been substantiated by the engineers who have used the geomembranes. Furthermore, actual construction cases reinforce the fact of the economic savings available from using a geomembrane. These studies, particularly the WyDOT horizontal and vertical placement projects, have shown costs that are in fact, lower specifically than the Texas unit prices for which the DVFMB is usually specified. The ease of the horizontal placement process makes the difference. The geomembranes have been used in over 100 U.S. projects, and they have been used successfully.

Information continues to be gathered in Australia with cooperative projects involving VICRoads and Swinburne University. The results have included the development of a trenching machine cutting 6 to 8 feet (2 to 2.5 m) deep with a width of only 3 to 4 inches (78 mm) minimum. The design was based on the 2-inch pitch chain and $\frac{1}{4}$-inch (6 mm) clearance between the chain and cutting edge of the trencher. These improvements will reduce the first cost of applying the geomembrane vertically. They will reduce the life-cycle cost for road maintenance expenses due to the geomembrane's ability to minimize damages caused by expansive soils.

## 11.6   References

1. M. B. Hager and G. Riedl. Unpublished data, 1993. (WYDOT)
2. Joni Kachelhoffer. Unpublished data, 1997. (WYDOT)
3. M. B. Hager. Unpublished data, 1997. (WYDOT)
4. Jim Miller. Unpublished data, 1997. (WYDOT)
5. L. Scofield and J. Lawson. Unpublished data. (Arizona DOT)

6. E. Cotecson and A. Myron. Unpublished data, 1993.
7. R. Garber and T. Yarger. Unpublished data. (Montana DOT)
8. R. Lueck and B. Forte. Unpublished data. (NMXDOT)
9. G. Browning. *Evaluation of Soil Moisture Barrier.* Mississippi Department of Transportation, Jackson, 1992.
10. A. J. Spensko. Unpublished data, 1993. (Utah DOT)
11. D. Cochran. Unpublished data, 1993. (Nevada DOT)
12. Arvid Noori. Unpublished data, 1997. (Nevada DOT)
13. Douglas Smith. Unpublished data, 1993. (Kentucky DOT)
14. E. B. McDonald, "Experimental Moisture Barrier & Waterproof Surface," South Dakota Department of Transportation, Pierre, 1973.
15. Daris Ornsbacher. Unpublished data, 1997. (S. Dakota DOT)
16. R. C. Gordon and T. J. Waters, "A Case Study of Performance on an Expansive Soil Subgrade," Fifth International Conference on Expansive Soils, Adelaide, 1984.
17. J. C. Holden. *Reduction of Pavement Damage from Expansive Soils Using Moisture Barriers.* Australia-New Zealand Conference on Geomechanics, Christchurch, New Zealand, 1992.
18. R. P. Evans and J. C. Holden. *The Use of Geomembranes to Control Pavement Movements from Expansive Soils.* 1992.
19. R. Jayatilaka, D. A. Gay, R. L. Lytton, and W. K. Wray. *Effectiveness of Controlling Pavement Roughness Due to Expansive Clays with Vertical Moisture Barriers.* Research Report 1165-2F. Texas Transportation Institute, Texas A&M University, College Station, 1993.
20. R. P. Evans, "The Stabilization of Pavements on Expansive Soils Using Vertical Moisture Barriers," Masters thesis, Swinburne University, Melbourne, 1995.
21. C. Mathias. Unpublished data, 1997. (So. Aus. TBO)
22. R. P. Evans, J. C. Holden, and K. McManus. "Application of a New Vertical Moisture Barrier Construction Method for Roadway Pavements." *The Management of Moisture in Expansive Soils,* Seminar, Swinburne University of Technology, April 1997.
23. S. Nesichi. Unpublished data, 1997. (Israel DPW)

# 12
# Case Studies: Airports, Canals, Railroads, and Pipelines

Transportation facilities other than highways, roads, and streets are also subject to the destructive movements of expansive soils. However, there is apparently less use of geomembranes in seeking to minimize these damages. This chapter discusses several cases in which expansive soil has created a hazard for airports, canals, railroads, and pipelines.

## 12.1 Airports

Significant studies have been conducted on U.S. airports in areas known to have expansive soils. Similar investigations have taken place in Israel and possibly Colombia. Gordon McKeen of the New Mexico Research Institute has conducted a thorough and extensive survey for the Federal Aviation Agency. Six airport sites in various states were identified as having been constructed over an expansive soil subgrade. A determination of the location of the zone of moisture activity at these sites indicated the depth to be 8 feet below the ground surface. This finding has lent credence to the recommendation that vertical placement of a geomembrane should be 8 feet deep.

McKeen noted that several airfields have significant runway distortions caused by expansive soils. However, the Dallas-Fort Worth International Airport has reported that it has not suffered any of these problems to date. The reason for their success may lie in their preventive measures. They designed and built their runways with 2 feet of lime treatment for the subgrade, 17 inches of cement-treated base, and 24 inches of concrete pavement. Current reports indicate that the runway pavements are not experiencing any distortions. However, apron and service areas that did not receive those substantial pavement sections are suffering some destructive expansive soil damages.

The recently completed Denver International Airport presents a study worthy of continued observation. Located northeast of the city, the airport was sited in a known expansive soil area. An internationally renowned Denver geotechnical engineer wrote a dozen letters to appropriate officials concerning the challenges these soils would present. The officials' only response was that they had had no such problems at the nearby Stapleton Airport. True, but Stapleton was built over a sandy subgrade.

One of the consulting engineers recommended to the officials that they use geomembranes in the runway assigned to his firm. The suggestion was not accepted. Instead, the final design called for the replacement of the expansive soil by nonswelling material. A cement-stabilized base and 19 inches of concrete pavement completed the runway construction. In 1996, a photograph in a Denver paper showed three engineers examining the cracks in the runway's concrete pavement. The airport paving engineer was quoted as saying that concrete always cracks. The geotechnical engineer indicated that they hadn't seen anything yet. Expansive soils would possibly claim another victim. Several lawsuits have been filed, but the court has sealed the documents, which has prevented public access to the facts of this case.

One use of geomembranes in airport construction is located in North Las Vegas, Nevada, two are in California, and apparently two are in Colombia, South America. The North Las Vegas airport used a geomembrane placed horizontally over the expansive clay subgrade. It is reported to be functioning well, and there is little runway distortion.

The Israeli airport study showed the effectiveness of an underdrain system. One edge of the runway had the underdrain while the other side had none, and it was found that the elevation changes were much less along the underdrain side.

## 12.2 Canals

Canals in the United States and Israel have been the focus of studies in expansive soils areas. In the 1950s the U.S. Bureau of Reclamation

(BUREC) was the first governmental agency to use geomembranes. The results were positive, and they have since used geomembranes in other places to control seepage in canals and dams and in reservoir rehabilitations. The BUREC uses the definition of *geomembrane* that has been adopted by industry worldwide to include such materials as synthetic linings, plastic linings, and flexible-membrane linings [1]. Since 1968 the BUREC has used 5 million square yards of geomembranes as canal liner, 1.7 million square yards as reservoir liner, and 100,000 square yards as dam rehabilitation liner. Forty-nine projects are listed in the report. Thicknesses used varied from 10 to 40 mils. Some of the geomembranes were covered with embankment material, some with shotcrete, and some with concrete. Other sections have been left uncovered. The geomembranes are generally placed from the top of the canal berm down the sides and across the bottom. The average unit costs on one of the projects was $1.70 per square yard for excavation, $.08 for preparing the subgrade, and $1.68 for furnishing and placing the 20-mil PVC geomembrane.

In *Remediation of Existing Canal Linings,* efforts to control seepage on the earth- or concrete-lined canals in the western United States are examined [2]. Polyvinyl chloride (PVC) was used because it remains flexible over a wide range of temperatures, is available in large panels, is easily field seamed, and has good puncture resistance and adhesion. It was used in 10-mil thicknesses on a Montana project in 1984. On a later project a 20-mil-thick geomembrane was used to achieve an increased service life. In later projects, polypropylene geomembranes were used for rehabilitation contracts. The major goal of these efforts is to control seepage losses, which in some cases have been as high as 50 percent. No significant note is made in the reports as to whether they are dealing with an expansive or nonexpansive soil. The geomembranes have been judged, however, to be significantly reducing the seepage.

The Bureau of Reclamation has used geomembranes as water barriers on dikes along the aqueduct between Scottsdale and Phoenix, Arizona. The Hayden Rhodes Aqueduct Reach 11 Flood Retention Dikes are of homogeneous silty and clayey earthen soil construction of about 15 miles in length. The dikes had cracked in three of the four locations, exhibiting settlement. Various alternatives were evaluated, and a high-density polyethylene (HDPE) was chosen for the water barrier. It was driven 50 feet down through the center of the dike. Frames were used for the placement of the geomembrane, which was 80 mils (0.03 mm) thick. ASTM specifications were used in specifying the performance requirement for thickness, density, and tensile properties. Factory seams and QA requirements were specified by BUREC. The report concludes that the geomembrane barrier has worked well [3].

In these reports the BUREC did not identify expansive soils as the problem they were contending with. In their early use of geomembranes at Cottonwood Dam, their intention was to reduce spillway erosion. To accomplish this, they used a HYPYLON geomembrane with a 12-inch subgrade cover. The BUREC's procedures in classifying soils works with Atterberg limits, gradation analyses, and their expansive uplift test. In their rehabilitation of the Nebraska canal and their repair work to the collapsed Tucson aqueduct, they used a very light density (VLD) geomembrane. At the Clotella Canal, a geomembrane lining has been used to protect the concrete from the negative impact of the high-sulfate soil.

The Friant Kern Canal in California's central valley has experienced slope failures in expansive soil areas, which has been addressed by flattening the slopes. Lime treatment of the subgrade walls has also been used with some remedial success. This canal is part of California's central valley project. As expansive soils are known to be present, it is interesting that no investigation into the use of geomembranes has been noted.

It has been reported that in a canal that carries water to Mexico City, a geomembrane has been used to control an expansive soil problem.

## 12.3 Railroads

China has used geomembranes to fight the destructive movements of expansive soils on their railway system. An estimated 20 percent of their system is built over expansive soils. These soils have caused considerable roadbed and rail distortion. Annual repair efforts have been costly. However, placement of horizontal geomembranes has yielded encouraging results. These expansive soil damages have been estimated at 100 million yuan (Chinese currency) annually. Geomembranes were used horizontally on the rehabilitation of a section of the Tai-Jao rail line in northern Shanxi Province. The subgrade was a dark brown silty clay with a grain composition of 50 percent silt and 40 percent clay. Its liquid limit was 47 percent, its plastic limit was 28 percent, and it had a free swell of 51 percent. The mineral composition of the soil was 12 percent montmorillonite and 18 percent illite. The soil has caused excessive lateral displacement, resulting in slopes' sliding into the longitudinal ditches paralleling the tracks. In addition, the soils have caused boiling and mud pumping, which has reduced the effective depth of ballast. This has resulted in track and roadbank displacements and distortions.

The remedial rehabilitation involved the use of soil piles and geomembranes and the repairing of the slopes and ditches. The cement-stabilized soil piles were 20 cm in diameter and 2.5 to 3.0 m (8 to 9.7 feet) deep. The track was raised, and ballast was removed on the section. The polypropylene fabric, 0.95 mm (40 mils) thick and coated with a bitumen, was placed between 5 and 7 cm of sand. The work was done in 25-m (80-foot) sections to minimize disruption of train movements. Displacement measurements were taken initially every 2 months. They averaged only 5 mm (0.25 inch), gradually decreasing and with no lateral displacements. Boiling and mud pumping, a previous regular occurrence, no longer took place. Previously high remedial maintenance costs ceased, and maintenance expenses were viewed as minimal [4].

The National French Railway Company (SNCF) has been using a reinforced asphaltic geomembrane to protect railway subgrades since 1972. As reported at the International Conference on Geomembranes in Denver [5], the remedial measures were selected based on the subgrade soil types. The reinforced asphaltic geomembrane, Coletanche HTP4, was designed for water tightening of the railway subgrades. The reinforcement consisted of a nonwoven fabric made of long-fiber polyester. The binder was blown asphalt, which is harder than a roadway asphalt and has a higher softening point. The panels developed permitted installation 12 feet wide and 24 feet long. Tests after 5 and 10 years in place have indicated that the reinforced geomembrane is serving well. As clayey soils have been identified at the sites, it is regrettable that the Atterberg limits have not been included in the tests. Since clayey soils have infiltrated the ballast, it would seem that the French have been addressing the expansive soil challenge although they have stated their purpose as only water tightening of the subgrade.

A technical manual produced by the U.S. Departments of the Army and the Air Force, *Engineering Use of Geotextiles,* July 28, 1995, indicates that there has been expansive soil impacts on the transportation infrastructure. It recognizes the challenges facing particularly railroad lines in the United States. To prevent damage, lime pressure injection and other treatment methods have been used on the subgrade. In addition, drainage ditches, underdrains, and geogrids have been used. However, as of this writing, no record of using geomembranes has been received.

Finally, according to communication received from Du Pont-Luxembourg, the Russian railroad system is presently using Du Pont's Typar to control expansive soils [6].

## 12.4 Pipelines

Pipelines buried in expansive soils have been dealt with and reported on in the United States and Israel. These lines have been moved and broken by swelling soils. One solution strategy has been to use a nonexpansive soil as a backfill material. This has helped prevent the swelling soil from creating forces sufficient to cause pipes to break. This method is as applicable to short pipelines as it is to long pipelines. In the Corpus Christi, Texas, area along the Gulf coast, utility pipelines for water and sewage have broken at various locations. This has caused additional movements of the expansive soils, resulting in damages to the adjacent homes. These utility lines were polypropylene pipes, which it was believed would not break under such conditions. Officials and engineers in the area are thus questioning the durability of this piping material.

## 12.5  References

1. G. Kassiff, M. Linneh, and G. Wiseman, "Pavements on Expansive Clays," Jerusalem, Academic Press, Jerusalem, 1969.
2. R. G. McKeen, "Design of Airport Pavements for Expansive Soils," Federal Aviation Agency, U.S. Department of Transportation, Washington, D.C., 1981.
3. J. Uzan, S. Frydman, and G. Wiseman, "Roughness of Airfield Pavement on Expansive Clay," Fifth International Conference on Expansive Soils, Adelaide, Australia, 1984.
4. M. Linneh and J. Ishai, "Israeli Experience with Runway Pavements on Expansive Clays," Sixth International Conference on Expansive Soils, New Delhi, 1987.
5. R. C. Rochlin, "Water Conveyance Pipelines in Expansive Soils," Fourth International Conference on Expansive Soils, Denver, 1980.
6. William R. Morrison and Alice I. Comer. *Use of Geomembranes on Bureau of Reclamation Canals, Reservoirs, and Dam Rehabilitation.* U.S. Department of the Interior, Bureau of Reclamation, Washington, D.C., 1995.
7. Alice I. Comer, Michael Kubt, and Kenneth Sayor, *Geotextiles and Geomembranes 14.* Elsevier, Amsterdam, 1996.
8. Jay Swilhart, Jack Haynes, and Alice F. Comer, "Deschutes-Canal Lining Demonstration Project"—Construction Report, Bureau of Reclamation, U.S. Department of the Interior, Denver, 1994.
9. M. Bliss and P. Brunette. "Reach 11 Dikes Modification, A Vertical Barrier Wall of HDPE Membrane," *Proceedings of the 1995 Geosynthetics Conference,* vol. I, Nashville, 1995.
10. X. M. Wu, P. X. Wei, and J. H. Chen. *Damage to Roadbeds and Remedial Measures.* Seventh International Conference on Expansive Soils, Dallas, 1992.
11. B. E. Hall, "Some Observations on the Performance of Railway Lines Over Expansive Clays" (South Africa), Seventh International Conference on Expansive Soils, Dallas, 1992.

12. Alain Le Coroller, Roland Savage, and Bernard Imbert. *Use of Asphaltic Geomembrane for Protecting Railway Subgrade.* International Conference on Geomembranes, Denver, 1994.
13. G. Hopp. Unpublished data, 1997.
14. R. M. Koerner and J. P. Welsh, "A Construction and Geotechnical Engineering Using Synthetic Fibers," John Wiley and Sons, New York, 1980.

# 13
# Placement

The placement of a geomembrane, whether horizontally or vertically, raises many questions. How do you do it, particularly a vertical placement? How is the trench for the vertical placement excavated? How is the backfill placed, and what material should it be? These are good questions, and they will be answered in this chapter.

## 13.1 Equipment Used in Remedial Treatments without geomembranes

Many different types of equipment have been used in efforts to control the destructive movements of expansive soils. The ponding techniques involve earthmoving equipment—dozers or scrapers—used to build dikes. An improvement in the ponding procedure has been made by the Mississippi Highway Department. They drilled holes 20 feet deep and filled them with a permeable material, which allowed ponding to occur. This required a drilling rig and a source of permeable material, along with the earthmoving equipment. Considerable success was achieved with this method on an interstate highway in the Jackson, Mississippi, area.

Another technique often used is to treat the expansive subgrade with lime, for which earthmoving and compaction equipment are needed. Laboratory tests provide the subgrade PI, its density, and the percent of lime needed. The equipment needed would include a maintainer and a traveling mixer to follow the trucks spreading the lime. Should off-site mixing be required due to the proximity of homes or businesses, vehicles would be used to haul the subgrade to a mixing site. There a batch plant would provide the mixed material, ready for return to the construction site.

Lime pressure injection has been used both in building site remediation and on transportation facilities. The equipment is frequently provided by a specialty contractor. The injection tubes are capable of placing the lime solution at a variety of depths in the subgrade. The lime slurry is also used for application on the job site subgrade treatment. Dalles Little's *Stabilization of Pavement Subgrades and Base Courses with Lime* gives detailed descriptions of these activities [1]. The pressure injection equipment has also been used to attempt to treat the swelling subgrade with patented chemical treatments. Arizona's DOT did some field testing to replicate Casagrande's electroosmotic chemical treatment of the expansive clays. However, they found the costs were high and that this treatment option was therefore not viable.

Asphalt has been used in several ways to control expansive soils. One is mixing the asphalt with the subgrade. This requires equipment similar to the lime mixing operation and to the cement stabilizing method. Cement and flyash additives have been used in a like manner. Another method of using asphalt has been to apply it heavily to the subgrade. The quantities used in the Houston, Texas, freeway subgrade encapsulation and by the Colorado DOT in a similar project were in the range of 1 gal/yd$^2$. For this operation an asphalt distributor is required along with the earthmoving equipment. A similar application was used by Mississippi DOT, who reported a year later that no sign of the asphalt remained. The poor results of the Colorado DOT tests with the heavy asphalt application led the Du Pont Company to offer them their Typar geomembrane for a free trial use.

Removing and replacing the expansive subgrade is another approach to meeting the challenge. Several projects have been tried removing 5 feet or more of the swelling soil and replacing it with a relatively inert substitute. Earth moving, compacting, and front-end loading equipment are needed. This approach has been used on several Texas highway projects, as well as in the construction of the Dallas Fort Worth and new Denver International airports. It seems to be working well in some cases. On the Texas IH 10 project in Hudspeth County, soil replacement presented several recurring questions: Does the less active replacement soil meet the required specification? Or did the more active material work its way upward through the select material? In the Texas case, the roadway later suffered substantial deformations, necessitating costly remedial work.

## 13.2 Horizontal Geomembrane Placement Methods

Placing the geomembrane horizontally provides the least placement challenge. On one of the earliest TexDOT projects, General McMullen Drive, THD maintenance forces unrolled the geomembrane by hand on the finished subgrade. Other means include placing the rolls on a piece of rubber-tired equipment and unrolling them as the vehicle moves along the subgrade. This equipment could be a tractor that is pulling a compactor or roller. The attachment to hold the geomembrane roll is a round bar, and side attachments to hold it to the tractor.

## 13.3 Vertical Geomembrane Placement Methods

Equipment designed to excavate a trench for vertical fabric placement is advancing rapidly in capability. In the beginning, however, contractors simply modified their existing equipment. A backhoe was used on the initial TexDOT project in the late 1970s. On the IH Loop 410 rehabilitation contract, the contractor's choice was a backhoe. The operation cut a wider trench than necessary, and the backhoe's hitting the trench walls being excavated probably contributed to the walls' sliding. When the subcontractor followed with another backhoe for the excavation, he used a sliding shoring pulled along by the backhoe, which solved the problem of the collapsing trench walls.

On the next TexDOT project, IH 37, the same subcontractor used a Vermeer trenching machine to cut the section for the fabric. The trencher cut to a narrower width than the backhoe. The expansive soil was in a stiffer condition, and no problem with sliding occurred. The daily goal of 400 feet production for the IH 410 project was frequently reached by 1 P.M. That was also the time the day's vertical fabric placement ended. Daily high production using trenchers, vertical fabric placement, and backfilling reached 2600 feet on later projects. Ditch Witch and Vermeer trenchers were also used as well as backhoes on other contracts.

On the U.S. 281 project in San Antonio, the contractor used a backhoe, and when prodded to increase production, he brought in a second backhoe. The choice of machine to do the work is usually the contractor's. In the work on IH 10 near the El Paso District's east boundary, the choice was to use several backhoes. The Wyoming contractors on their state's

DOT contracts have generally used the Ditch Witch trenchers to handle the excavation.

Vertical geomembrane placement has required a variety of devices. They usually include a rubber-tired dispenser that holds the roll horizontally parallel to the trench centerline. The fabric is then turned on a bar, which passes it down into the trench. The coated side is often placed against the trench wall to avoid any damage from the backfilling operation.

## 13.4 Geomembrane Trench Backfill Methods

Methods used in the placement of the backfill material have varied according to the material used. The IH 410 project called for a sand backfill. The trench, the width of which was not specified, was 2 feet wide, and the sand was bladed into the area. No special compaction methods were required. Within a few months, two vehicles became stuck in the vertical barriers' trench sand backfill. One solution planned to prevent future problems was to add cement to the top foot of the backfill. Before this plan could be implemented, however, reports of vehicles' getting stuck in the sand ceased. Consequently, no further steps were taken.

A washed concrete aggregate was used on the next DVFMB project. This mixture was bladed into the trench after an underdrain pipe was placed. A cap of cement-stabilized base, 2 feet thick, followed. No problems were reported from this solution. The cement-stabilized base cap was poured into the trench from a tremie on the ready-mix truck. The IH 10 project in Bexar County, Texas, between FM 1516 and the Guadalupe County line called for using a coarse-base screening backfill. When this source was expended, the contractor received permission to use a finer-base screening material. This material was used without additional compaction operations, and later the sections were found to be suffering from voids permitting shoulder cracking and displacement. Corrective action followed by grouting the areas suffering the voids, and no subsequent difficulties have been reported. The grout was pumped into the trench at the needed locations.

The project on U.S. 84 near Snyder, Texas, used a variety of backfill materials including some excavated from the trench. The indications so far are that varying material used for the backfill made little difference. On this project the contractor devised a wheel compactor for the backfill material. It didn't work well, and its use was discontinued. WyDOT has used the material excavated from the geomembrane's trench as backfill material with similar success. Again the contractors frequently used homemade wheel compactors for the backfill operation.

## 13.5 Australian Backfill Methods

The success of recent Australian methods might substantially alter backfill placement equipment requirements. The developments are explained in two recent publications, Tech Note 13, *Vertical Moisture Barriers* (VICRoads, Materials Technology Department, October 1995), and *Application of a New Vertical Moisture Barrier Construction Method for Highway Pavements* by R. P. Evans, J. C. Holden, and K. J. McManus [2]. With the considerable areas of the Australian states of Victoria and South Australia having expansive soils, the vertical moisture barriers have been found to be a desirable solution to the challenge, and their method is less expensive than the Texas approach. The Australians concluded that the use of gravel for backfill in a 2- to 3-foot wide trench was not a viable option as it is possibly a carrier of water to the bottom of the excavation. They addressed the problem by first determining that the moisture reduction would amount to 90 percent with a 6-foot (2-m) barrier compared to 100 percent for a 9-foot (3-m) barrier.

The Australians modified a trencher to make a 6-foot (2-m) with a 3-inch (78-mm) wide cut. A slim-line trenching boom was designed and built, as was a crumber bar to keep the bottom of the trench free from spoil. A soil chute was attached to the trencher to take the spoil to a conveyor belt to be deposited away from the trench. Along with the narrower trench, they developed a flowable backfill of a medium-graded sand, a low cement content, high flyash proportion, and water content. The flowable backfill is poured into the narrow-slit trench using a specially designed mobile hopper from a ready-mix truck. A guiding roller suspended from the hopper into the trench ensures that the geomembrane will not be damaged. Poured directly into the trench, the backfill has minimal labor requirements and is self-compacting. The result is a cost reduction from A$35 (Australian dollars) per 9 feet (3 m) for the same depth with a 1-foot (300-mm) wide trench to A$15 for the narrower trench, the geomembrane, and the flowable backfill.

The geomembrane used is a double layer of thin, 0.2-mm, low-density polythene, which is cheap and easily available in Australia. A single layer of the material was used on an earlier project, but it was found to suffer tears from construction damage and tree root penetration. The double layer is tougher and less likely to be damaged. The geomembrane is placed using a specially designed dispenser, carried horizontally, and then it is vertically passed around a bar at a 45 degree angle. Its placement is on the shoulder side of the trench rather than the usual Texas pavement side placement. On their Merwyn Swamp project in northwest Victoria, a production rate of 180 feet (60 m) per hour was achieved [2]. For a 10-hour day, this could reach 1800 feet (600 m), which is quite good. Their modifications are proving very successful and merit close observation.

## 13.6  References

1. W. G. Watt and M. L. Steinberg, "Measurements of a Swelling Clay in a Ponded Cut," Research Project 118-6 Center for Highway Research, Austin, 1972.
2. P. T. C. Teng, R. M. Mattox, and M. B. Clisby, "A Study of Active Clays as Related to Highway Design," Mississippi Highway Dept., Jackson, 1972.
3. M. L. Steinberg, "Horizontal Placement of a Geotextile on a Subgrade to Control a Swelling Soil," Research Report 187-9, Center for Highway Research, Austin, 1983.
4. M. L. Steinberg, "Deep Vertical Fabric Moisture Seals," Fourth International Conference on Expansive Soils," Denver, 1980.
5. M. L. Steinberg, "Controlling Expansive Soils: Twenty Texas Highway Projects," Seventh International Conference on Expansive Soils, Dallas, 1992.
6. G. Browning, "Evaluation of Soil Moisture Barrier Construction," Mississippi Department of Transportation, Jackson, 1992.
7. J. C. Holden, "Reduction of Pavement Damage from Expansive Soils Using Moisture Barriers," Sixth Australian-New Zealand Geomembranes Conference, Christchurch, New Zealand, 1992.
8. R. P. Evans, J. C. Holden, and K. J. McManus. *Application of a New Vertical Moisture Barrier Construction Method for Highway Pavements.* Paper presented at the Eighteenth ARRB Transport Research Conference, Christchurch, New Zealand, September 2–6, 1996.
9. Dalles Little. *Stabilization of Pavement Subgrades and Base Courses with Lime.* National Lime Association. Kendall Hunt, Ames, Iowa, 1995.

# 14
# Costs

Cost considerations are critical elements in any engineering decision, be it design, construction, or rehabilitation. The long-range costs of not planning adequately for the destructive forces of expansive soils are becoming better known and of greater magnitude. The costs of using geomembranes, however, have declined considerably in the past two decades. This cost reduction has enhanced the level of comfort in placing geomembranes.

Geomembrane specifications set limits and requirements that determine the boundaries of the type and price of the geomembrane to be used. Supplementary specifications should include such items as backfill details, cap treatment, and underdrain pipes as well as workload and the availability of workers and equipment to do the job. Reports of Australian developments are presented with their impact on the cost of placing vertical fabric moisture barriers. In their reports, they consider placement in relation to the costs of other types of treatment.

## 14.1 Factors Affecting Costs

Variation in the costs of using geomembranes depend on the specifications related to thickness and other requirements. As the thickness requirement, as expressed in mils or ounces, increases, the geomembrane costs also increase. The polypropylene or other hydrocarbon-based fabric will tend to give a lower price for thinner, lighter engineering fabric. Current prices tend to vary from $0.59 to $1.35 per square yard.

## 14.2 Cost Variations on Texas Highway Projects

Cost variations over the years on Texas highway projects reflect a range of factors. The significant trend seems to be a considerable decrease in construction unit bid prices over the decades. The first Texas highway geomembrane project took place in the later 1970s on General McMullen Drive. The geomembrane was supplied at no charge by the manufacturer and installed by state maintenance forces. The next project using geomembranes as a deep vertical fabric moisture barrier (DVFMB) was on IH 410 in the Valley Hi Drive Underpass area. The contractor's bid price was $20 per linear foot for supplying the geomembrane, Du Pont Typar T 063, excavating to place it to a depth of 8 feet, tacking the geomembrane 2 feet to the paved shoulder, and backfilling the trench with sand. This price was close to the $21 per square yard bid on IH 37 the next DVFMB project. The bid per foot on IH 410 was made for a length along the roadway of the geomembrane placed 8 feet deep and tacked 2 feet to the paved shoulder. Since the geomembrane came in a 3-m width, this was considered close to the square yard estimate. Some differences included an underdrain pipe placed at the trench bottom and a cap consisting of the gravel backfill material of 2 feet of cement-stabilized base.

Variations appear in bid prices tending to obscure several factors. On the U.S. 281 rehabilitation contract, the DVFMB was bid at $3.18 per square yard. This price might indicate that the cost of the DVFMB was shifted to another contract bid item. For a 1985 rehabilitation contract on IH 10 in east-central San Antonio, Texas, the DVFMB was bid at $15 per square yard. For another San Antonio area project on U.S. 87, the geomembrane was bid at $13.50 per linear foot. For several IH 10 rehabilitation contracts east of San Antonio, the DVFMB was bid in the $13 per linear foot range as was another project on IH 10 in far west Texas. Bid prices sometimes get massaged for different reasons possibly including getting prepayments up front to contractors. In the largest Texas contract involving DVFMB on IH 45, the geomembrane-in-place bid price was $6 per linear foot. Over $1 million of geomembranes were placed on this contract, and the large project quantity had to be a positive factor keeping the price down.

One noteworthy estimate was made by an engineer from an Australian trenching machine manufacturer on a Texas sales trip. Working with an estimated geomembrane price of $1.25 per square yard, placing it 8 feet deep in a trench, adding backfilling and factoring in profit, the bid projected was $5 per linear foot.

Costs

**Table 14-1.** WYDOT Impermeable Plastic Summary

| Year | Quantity, square yards | Projects | Average price per square yard |
|---|---|---|---|
| 1987 | 58,120 | 2 | $2.33 |
| 1988 | 9,900 | 1 | $2.31 |
| 1989 | 191,000 | 3 | $1.28 |
| 1990 | 622,100 | 6 | $1.38 |
| 1991 | 418,100 | 4 | $1.30 |
| 1992 | 1,135,990 | 10 | $0.71 |
| 1993 | 891,000 | 10 | $1.10 |
| 1994 | 324,000 | 5 | $1.01 |
| 1995 | 221,350 | 5 | $1.21 |
| 1996 | 316,582 | 6 | $1.22 |
| Totals | 4,188,142 | | |

While these reductions in bid prices on Texas highway projects are substantial, $20 to $6 per linear foot, they reflect placing the geomembrane 8 feet deep. The work by the Wyoming Department of Transportation reflects their usual placement, horizontally and vertically. Since the majority of the geomembrane is laid horizontally and the vertical depth varies from 3 to 5 feet, the price reflects to a degree what the plans and specifications require. A recent tabulation, Table 14-1, shows two geomembrane projects going to contract in 1987 using 54,120 square yards for an average price of $2.33. In 1992 a total of 10 projects with geomembranes went to contract totaling 1,135,990 square yards. The average bid price was $0.71 per square yard. The 1986 price for 6 geomembrane projects was $1.22 per square yard.

## 14.3 Accuracy of Project Specifications

The accuracy of the cost projections will depend on the thoroughness of the specifications. Careful consideration must be given to the vehicle loads the geomembrane will face along with the type of equipment. Also, an awareness of the record of a variety of similar projects will provide considerable help in estimating costs. From the earliest Texas projects, on General McMullen Drive and IH 410, the possibility of the geomembrane's tearing has been a matter of concern. In neither of these projects, however, did the geomembranes tear, although they were no more than 20 mils thick. The weights varied from 2.5 to 7.5 oz/yd$^2$. On the General McMullen Drive work, the base was dumped on the geomembrane from trailer trucks. The base was knocked down and laid with a maintainer, sometimes called a *road grader*, and water trucks and rollers used in the process. No damage to the geomembrane was visible.

In other situations where the subgrade has sharp-edged rocks embedded in the clay, a stouter geomembrane should be considered. No report has been made by any of the geomembrane sites where it has been used to control expansive soils indicating the material has been torn. The conclusion is the 10- to 20-mil-thickness, weighing from 2.5 to 9.0 oz/yd$^2$, serves very well.

## 14.4 Comparison of Australian and U.S. Costs

A comparison of Australian and U.S. assessments has been the impetus for some important cost savings. An evaluation of the U.S. $6 bid price on the Texas IH 45 project near Waxahachie, which called for the use of a 3-m (9.75-foot) deep trench that was 300 mm wide (12 inches) with a comparable U.S. geomembrane, formerly a Phillips Petromat MBII, now an Amoco product, would cost A$27.30 (Australian dollars) per linear meter. Evans, in his master's thesis, projects that with Australian materials and labor, the cost of the "Texas" method would be A$45 per linear meter. This would assume using the excavated spoil for backfill. Had another backfill material been used with a cement-stabilized cap, the cost would have risen to A$52.70 per linear meter. The MBII geomembrane is on the heavier end of the material spectrum and would be used at a rate of 9 oz/yd$^2$. The cost in Australia was estimated to be $3.75 per square yard. The total cost of the Texas method was considered possibly too expensive for the local resources.

A more economical and more effective barrier design has been developed by the Australians. Its aim is to reduce, and whenever possible, eliminate the labor cost. A flowable backfill was found to be the most expensive single element of the barrier design. To minimize this cost, the trench width was reduced. The width reduction also affected the trench depth and the chain's size or pitch. They related these elements to the size of the trenching machine to provide the power, the required rate of production, and the soil type. They developed a table to show this interrelationship:

| Chain pitch, in | Practical minimum trenching width | Maximum excavation depth |
|---|---|---|
| 1⅝ | 65 mm (2¾ in) | 1.5 m (4.8 ft) |
| 2 | 78–80 mm (3¼ in) | 2.25 m (7.5 ft) |
| 3 | 100 mm (4 in) | 3.0 m (9.75 ft) |

The chain pitch sizes are standard. Two thorough detailed cost comparisons are presented in Evans's thesis. One is the cost of constructing the new moisture barrier using the Australian method in Doonen, Australia. The second is the cost of the same project using the Texas method. These studies are based on 0.5-m increments up to 3 m. The new Australian method costs varied from $4.14 to $34.50 per meter for 0.5 to 3 m. For the Texas method, the same costs ranged $9.00 to $20.40 for 3 m, if the excavated spoil is used for backfill. For the Texas method, the 3-m cost with a pea gravel or other special backfill would reach A$52.70 per meter. A point to remember is that the trench width in Texas is not a required 1-ft width nor is pea gravel required as a backfill. Those decisions are left to the contractor, who must make the most economical combination that still meets the specification requirements.

As Evans describes in his thesis, specifications were developed for the planned thin-line trencher. The power required would enable trenching 2 m (6.5 feet) deep in stiff clays at a speed of 50 m (150 feet) an hour. The trenching teeth would excavate a minimum width, and the sides of the boom would be as smooth as possible to minimize frictional drag along the trench wall. A slim-line crumber bar was needed to remove spall from the bottom of the trench, and at the other end a conveyor belt would deliver the excavation away from the trench. This would mean that the operation of the fabric dispenser and the flowable backfill could operate more closely to the trench. A soil ramp was needed to deliver the excavated trench material to the conveyor belt, and several booms were also needed. Furthermore, the boom design would need to be easily adaptable to a wide range of trenching machines.

The trenching machine investigation involved the products of six manufacturers. Included were Case, Ditch Witch, Perry, Steenbergen Hollandrain, Yemec, and Vermeer. To excavate the 2 m (6.5 feet) a 40- to 50-horsepower machine was necessary. An available trenching machine was fitted with a 2-inch pitch chain that could cut at a 68-mm width. Tungsten chisel teeth were used, supplied by Pengo, Model 1336 TC, to cut at 80 mm (3+ in) with the 2-inch chain. Teeth placement on the chain became a factor to consider, as did the bottom idler wheel on the boom.

Another important consideration was that a maximum tension would have to be kept on the boom and the bottom of the trench would have to be kept clean. A spring-loaded crumber bar could be used to keep the trench bottom clear, and this would also help keep the trench width to a minimum. VICRoads asked a reputable manufacturer to help develop a trenching machine that could meet these requirements. Myrtleford accepted the challenge; when they felt confident that they had met it, they conducted a field test of the new manufactured equipment. In the test with a dry brown loamy material, a rate of 2.5 m (8 feet) per minute was reached [150 m (480 feet) per hour]. With the improved narrow trencher, additional economies would evolve.

Further investigation led the Australians to conclude that the seasonal moisture change zone was close to 2 to 2.5 m deep. Using the narrower trench excavator, the flowable backfill, and the polythene material, the Australian cost was reduced to A$15.17 per meter for the 2-m (6.5-foot) depth, or A$23.60 per meter for the 2.5-m (7.5-foot) depth. This would compare more favorably with the Texas method costs of A$15.60 per meter and A$16.00 per meter for the 2- to 2.5-m depths if an excavated spoil backfill were used. If a backfill of another material were used, Texas method costs in Australia would range from $35.55 to $44.15 per meter. The Australian innovations are most praiseworthy. Anything reducing the cost of creating the moisture barrier and reducing the volumetric changes of the expansive soil is a welcome improvement for widespread use [1].

## 14.5 Costs of Various Geomembrane Placement Rates

Placement production rates with variations noted and recognition granted for costs related to labor will be key elements in computing costs for a DVFMB. Contrast the 600-foot-a-day goal with placements of 2500 feet and its impact on labor costs. Placement production rates also impact schedules and the amount of time it takes for withheld funds to be released on work completion. The production rate is affected most by how the geomembrane is placed. In Wyoming the horizontal and vertical layout of the geomembrane results in an average daily rate of 3000 feet. This rate is further limited by the dumping of a base course on the horizontally placed fabric to protect it from nighttime traffic and the sun's ultraviolet rays during the day. These rates are achievable not only in the United States. Australia has reported 1-day DVFMB completion rates of 1800 feet.

## 14.6 Cost of Various Treatment Methods

Comparisons of various types of treatment are important costing considerations. Least cost and life cost are matters that need careful thought and evaluation. Sometimes first cost clouds the issue. In other instances a political situation may motivate the assessment of comparing different solutions. For example, Dallas Little has offered a complete methodology for calculating lime treatment costs [2]. Lime has been used extensively for centuries. In this century lime has been used since the early 1940s and was based frequently on the work of Conard Kelly. Lime treating 6 inches of the finished subgrade became a byword on Texas and Oklahoma highways in expansive soils areas. When in some cases this treatment didn't suffice and the subgrade continued to swell, lime treatment increased to 12 and 24 inches. First cost aside, the question remained as to whether it would do the job.

Little presents a thorough examination of life-cycle costing. He examines, among a variety of other items, agency costs, user costs, and nonuser (excluding agency costs) costs. Included are constant dollar studies, discount rates, and current dollar studies. It discusses the difficulty in predicting inflation rates, whether there will always be inflation, the effect of federal programs, debtors' gains by repaying loans with inflated dollars and should that occur, how should those benefits be measured, and a programs bias toward long-lived, capital-intensive projects.

It further presents the Recommended Analysis Life for Comparing Pavement Life, a tabulation by Epps et al:

| Activity | Pavement surface type | Recommended analysis life year |
|---|---|---|
| New construction | PCC only | 45 |
| Reconstruction | HMA only | 30 |
| Thick overlays | PCC and HNA | 45 |
| Rehabilitation | PCC only | 20 |
|  | HMA only | 20 |
| Maintenance | PCC only | 20 |
|  | HMA only | 20 |

Salvage values and life of rehabilitation alternatives are included in the presentation of life-cycle costing. The first is noted as being generally accepted. The second is taken from the U.S. Army Corps of Engineers' values of 1987:

| Pavement type | Typical life cycles representative range, years | Representative average, years |
|---|---|---|
| New PCC | 16–25 | 20 |
| PCC overlay | 10–20 | 15 |
| New HMA | 12–16 | 14 |
| HMA overlay | 5–15 | 10 |

The Corps of Engineers' tabulation lists 14 items, each with an average mean life cycle in years, average range of the life cycle, and the number of data points. The 14 range from crack sealing to patching to mud jacking. Average life-cycle times range from 3 to 4 years for the sealing and patching to 16 years for the mud jacking and 20 for slab replacement. Should additional data formulas and tabulations be needed, they are available in Little's book.

The cost-effectiveness of several lime treatment techniques are presented in Little's book as well in John Nelson and Debora Hamburg Miller's [3]. Lime and lime slurry pressure injection are prominently discussed, more often in the remediation of residences. Other chemical injection methods have been used to address this challenge. Care needs to be taken to study the past record of the cost-effectiveness of these treatments.

Looking at structure foundations, the drilled shafts developed by Willard Simpson, Sr., in the 1930s continue to be widely used. Fu Hua Chen mentions in his books, previously noted, that the shaft technique has been used with considerable success. Some problems have arisen with the drilled shaft in some structures, both large and small. These developments need to be carefully considered. PTI design recommendations for slabs on grade have been used often with success. Both in the United States and in Australia, geomembranes have been used to counteract the swelling action of soils on structures. Professor Kent Wray, formerly of Texas Tech University, now dean of engineering at Ohio University, and Kirby Meyer of Austin, Texas, a geotechnical engineering consultant, have used geomembranes in residential work with considerable success.

In highway design studies, replacement of the expansive soils has been used. The approach frequently has been to take out an additional 5 feet of the expansive soil and replace it with a nonexpansive soil. In several recorded cases, this has not worked well. On IH 10 in Hudspeth County in west Texas, pavement distortion reappeared after several years.

The undercut-and-replace approach is currently being used on a recently awarded contract on IH 45 south of Waxahachie, Texas. This section is adjacent to the IH 45 contract where over a million dollars' worth of DVFMB was placed. The costs of the recent contract are discussed in greater depth in Chapter 9 dealing with highways, roads, and streets. The bid price covering removing high-PVR clays from 36 to 52 inches is $1.92 per cubic yard. The price bid for the select fill to replace the more expansive material is $5.11 per cubic yard. Concern has been voiced about the success of the remove-and-replace operations on other projects. The prices for these earthwork items are frequently dependent on the type of equipment available for the physical constraints of the work site.

A bottom line for geomembrane costs is they have come down considerably since the 1970s. Construction techniques have improved. In addition, the Australians have developed the slim-line trencher with the flowable backfill, which could translate into a further reduction in cost and an increase in production rates.

## 14.7  References

1. M. L. Steinberg, "Deep Vertical Fabric Moisture Barriers in Swelling Soils," Transportation Research Record 790, Washington, D.C., 1980.
2. M. L. Steinberg, "Further Monitoring of Twelve Geomembrane Sites in Texas", State Department of Highways and Public Transportation," Austin, 1988.
3. H. Stanford, unpublished data.
4. G. M. Hager, unpublished data.
5. M. L. Steinberg, "Vertical Moisture Barrier Update," Transportation Research Record #136, Transportation Research Board, Washington, D.C., 1992.
6. R. P. Evans, J. C. Holden, and K. J. McManus, "An Improved Method for Installing Vertical Moisture Barriers," 8th REAAA, Conference, Taipei, 1995.
7. D. N. Little. *Handbook for Stabilization of Pavement Subgrades and Base Courses with Lime.* Kendall Hunt, Dubuque, 1995.
8. J. D. Nelson and D. J. Hamburg. *Expansive Soils, Problems and Practice in Foundation and Pavement Engineering.* Wiley, New York, 1992.
9. R. P. Evans. *The Stabilization of Pavements on Expansive Soils Using Vertical Moisture Barriers.* Master's thesis, School of Civil Engineering and Building, Swinburne University of Technology, Australia, September 1995.

# 15
# Durability

Durability is a matter of great concern when recommending any material or treatment for preventing expansive soil damage. A geomembrane is not forever—like most materials it can be destroyed and made useless. On the other hand, there are some simple measures that can be taken so that they will be able to serve their purpose well for decades. This chapter reviews what has been done to test its durability and to enhance its longevity.

## 15.1  Justifiable Concerns

Justifiable concerns about the geomembranes' longevity and durability have been addressed through many projects and studies. *Durability* refers to the long-term ability of the buried geotextile to resist degradation. It excludes polymer attack by ultraviolet light. Degradation in the ground can come from chemical, biological, or mechanical sources. There is substantial geotechnical literature in this area. Both longevity and durability relate to the survivability of the geotextile. The *Federal Highway Administration* (FHWA) *Geotextile Engineering Manual* defines *survivability* as its resistance to destruction during its initial road construction and operation phases [1]. These periods are considered critical for a geotextile. The FHWA provides tables to determine survivability rates for the construction phase. The more severe the site subgrade condition, the more uneven, the more severe the rating. The survivability ratings range from low to moderate to high and very high for the most severe installation conditions. Halliburton at Oklahoma State University did considerable work in studying this critical phase. He found that the inservice period following construction is less fraught with destructive dangers.

Paul Teng, then Mississippi Highway Department research engineer, raised a very good question about geomembranes: If heavy asphalt applications to a subgrade can be gone without a trace a year later, what will happen to an engineering fabric made primarily of carbon, hydrogen, and oxygen atoms? Why invest money in something that won't last any definable time period? Fortunately, research has put these fears to rest.

## 15.2 Tests by Manufacturers

Tests by manufacturers are frequently the initial steps to determining the basic properties of a material. In a recent paper, funded in part by REEMAY, a manufacturer, the author focuses on construction survivability and durability of two specific geotextiles used in road construction [2]. Both geotextiles—polypropylene, continuous filament, thermally bonded nonweaves—were exhumed from seven existing unpaved road surfaces. The sites ranged from the Texas Gulf coast to Washington state and to Illinois. The geotextiles weighed from 4 to 6 oz/yd$^2$. The field investigations and test program assessed the geotextile's construction survivability and long-term durability characteristics. Measured strengths after 1 to 12 years of burial ranged from 50 to 90 percent of their initial values. Loss of strength was found to vary with the severity of construction conditions. The strength loss was concluded to be caused by mechanical damage to the geotextile structure. Further sophisticated testing indicated no significant polymer degradation.

Another field test was reported by REEMAY and Exxon Chemical Company [3]. The farm road in Delaware had a 4-oz/yd$^2$ Typar 3401 fabric, developed by the Du Pont Company, placed on the subgrade, followed by 15 cm (6 inches) of gravel base. The soil was a sandy clay with a CBR of 1 when wet and 6 when dry. The geotextile section was 310 m (1000 feet) long with adjacent control sections having 38 cm (16 inches) of gravel base but no geotextile. The idea was to accelerate or encourage failure, intentionally underdesigned, then to evaluate the testing of the geotextile. Three other geotextile test sites had been established at the same time. The Smyrna site was the only one where the designer had project control.

Another construction test was conducted with the same gravel base thickness on the geotextile and no fabric sections. Fully loaded vehicles created soft spots after 120 passes over the Typar section. After only 29 passes, there was complete failure of the section without the geotextile. Twenty years later the site, which had been actively used, was uncovered and examined, and a sample of the geomembrane was taken to the labo-

Durability    **167**

ratory for further testing. Visual inspection of the roadway indicated rutted surface where there had been no geotextile. Where the Typar had been placed, there were no ruts. Shelby tube tests indicated that the geotextile had separated the subgrade from the base. The sections without the geotextile had significant intermixing in the ruts and potholes. The laboratory tests indicated that after 20 years, the geotextiles' physical characteristics had remained stable in their underdesigned roadway section.

Photomicrographs of the polypropylene filaments showed no degradation over time. The only damage appeared to be mechanical to the geotextile's outer layer. Differential scanning calirometry (DSC) testing resulted in the conclusion that the polymer degradation was minimal to nonexistent. The study concludes that the thermally spun-bonded polypropylene geotextile, Typar 3401, performed its design assignment of permeable separation on the Smyrna project. Visual inspection has shown that the material increases the performance of the unpaved road. Qualitative analysis has indicated that there has been no degradation of the Typar over the 20-year period.

## 15.3  Tests by State Agencies

A report issued by the Wyoming Department of Transportation (WyDOT) examines the durability of the geomembranes in considerable detail [4]. Twelve sites were sampled for follow-up studies where geomembranes had been placed. The purpose of the study was to determine if the geomembranes were doing what their design had intended. Typically, sections of geomembrane had been placed horizontally across the subgrade or subbase and then keyed vertically 3 to 4 feet (0.9 to 1.2 m). WyDOT first used geomembranes in 1981. A 1997 communication from WyDOT indicated between 1987 and 1996, 4,188,142 square yards of geomembranes had been placed on their highways. Their construction methods reflect their awareness that the geomembrane needed to be protected from damage by the sun's ultraviolet rays. The base covered the fabric placed horizontally, and no further fabric placement took place beyond the day's limit to cover it with base.

The 12 sites WyDOT investigated were located in different parts of the state. At each, a 2- by 2-ft (0.6- by 0.6-m) piece of membrane was exhumed. The samples were tested for strength and permeability. In addition, ten soil sample holes were drilled at each site. These samples were tested for moisture content, soil classification, and shear strength. The study reported that damage was seen in the exhumed geomembrane where holes had been punched usually by angular subgrade particles. It was conjectured that the damage had not been caused by the base placed on top of the geotextile. There were no signs of any unusual distress in the pavement section above the geomembranes. Laboratory testing included strength procedures outlined in ASTM D 4695. Elongation at failure was also determined in this test. Second test runs included ASTM D 4492 to determine permeability.

Strength tests of the samples generally indicated a decrease of almost 30 percent from the levels specified at the time of construction. There was no relation between the length of time in place to the decrease of strength. Average loss of elongation was 36 percent. The roadway surfaces remained in good shape despite these decreases from the design levels of strength and elongation. Though the maximum time the geomembrane had been in place was 5 years, this method of minimizing destructive movements of swelling soils has been considered a success, and the WyDOT continues to use the geomembranes.

Though the Texas DOT has not issued a report as comprehensive as WyDOT's, some testing and follow-up does exist, with positive results. Resident engineers for TexDOT in the San Antonio and Seguin offices have noted substantial decreases in maintenance expenses on IH 10, IH 410, and IH 37. On the last-named project, a 2-mile section had required from $50,000 to $100,000 in pavement maintenance expenditures annually. For the 12 years following the removal of the median ditch and the installation of a DVFMB, no pavement maintenance expenditures were required. On a section of IH 10 on San Antonio's east side, a similar condition of high annual pavement maintenance costs preceding a rehabilitation with a DVFMB was followed by a prolonged period in which no funding was needed. The resident engineer's quote is remembered: "The geomembrane made a believer of me." The durability of a geomembrane is seen in the results, the proof of the pudding, as it were. On the IH 10 project, ditch wall sliding took place. To remedy the situation, the trench for the geomembrane was moved farther away from the shoulder. The geomembrane was then laid horizontally from the relocated trench to the paved shoulder. To avoid exposure to ultraviolet rays, this horizontal placement was promptly covered by subgrade to protect it from degradation.

## 15.4 Australian Experiences

In Australia, 20 percent of its area is affected by expansive soils. These areas are largely found in the most heavy populated parts of their continent. In 1985 VicRoads began an experiment with the use of a vertical fabric moisture barrier. Two goals shaped their efforts: One was to reduce the destructive pavement movements caused by the expansive soil, and the other was to stop the roots of an adjacent tree grove from penetrating the roadway subgrade [5 and 6]. Their conclusions from the experiment were that the geotextile placed in 1986 had survived and was in good condition 7 years later. They further noted that the roots had been deflected parallel to the fabric and that they had not penetrated and entered the roadway subgrade. The road surfaces showed less distortion when the geomembrane was placed 2.5 m (8 feet) rather than 1.5 m (4.5 feet) deep. A later study substantiated the earlier results. The Australians found that the geomembranes had considerable durability. Their report also discussed the positive results of the use of a narrow trench excavator and a flowable backfill and their effectiveness in protecting buildings. In addition, the report notes the durability of the 1-mm, high-density polyethylene (HDPE) fabric. The flowable backfill was noted also to have a positive impact on durability as it is less likely to tear the fabric than a gravel or sand backfill [7].

## 15.5 Positive Indications of Geomembrane Durability

The manufacturers' testing gives us the first indication that the geotextile is a durable product. Although it does need protection from the ultraviolet rays of the sun, when so protected, it can safely be expected to last a significant amount of time. The tests reported by Bonaparte and by Hawkins and DeBerardino give strong indications that a geotextile, when properly coated, can create a geomembrane that has a significant life span of over 20 years.

The results reported in the WyDOT study give further credence to the expectation that the geomembrane will be wherever it is placed for a significant time period. The 20-year span reported by Hawkins and DeBerardino may, in fact, be only a small part of the geomembrane's life span. The test results achieved by TexDOT, who realized significant reductions in pavement maintenance expenses, give further indications that the geomembrane, whether used in vertical or horizontal moisture barriers, will be in place for decades. The Bureau of Reclamation's studies confirm the results of other tests as well. Apparently the first in the United States to use the geomembrane, the bureau continues to do so because they have achieved the results they were looking for.

## 15.6 References

1. B. R. Christopher and R. D. Holtz. *Federal Highway Administration Geotextile Engineering Manual*, 1984.

2. T. C. P. Teng and M. B. Clisby, "Experimental Work for Active Clays in Mississippi," Transportation Journal, ASCE, New York, 1975.
3. R. Bonaparte. "Survivability and Durability of a Nonwoven Geotextile." *Geosynthetics for Soil Improvement.* ASCE National Convention, Nashville, May 1988.
4. S. J. DeBerardino and W. M. Hawkins. *A Durability Study of a Typar Twenty Years After Installation: The Smyrna Road Project.* Paper presented at the Transportation Research Board 73rd Annual Meeting, Washington, D.C., 1984.
5. M. A. Falk and G. M. Hager. *Performance of Impermeable Membranes to Treat Swelling Soils: The Wyoming Experience.* Wyoming Department of Transportation (FHWA/WY 94/05), Cheyenne, 1994.
6. R. E. Magers, unpublished data.
7. D. R. Stein, unpublished data.
8. J. C. Holden. *Reduction of Pavement Damage from Expansive Soils Using Moisture Barriers.* Australia-New Zealand Conference on Geomechanics, Christchurch, New Zealand, February 1992.
9. R. P. Evans and J. C. Holden. *Use of Geomembranes to Control Pavement Movements from Expansive Soils.* UTS Seminar on Geosynthetics in Road Engineering, Sydney, Australia, September 1994.
10. J. C. Holden. *Some Developments in Urban Root Barriers.* Australia-New Zealand Conference on Geotechnics, Adelaide, Australia, July, 1996.

# 16
# Last, But Definitely Not Least

This is the windup. It's the last chapter in this book. So far, we have provided you with a great deal of valuable and useful information. Some you knew. Dealing with expansive soils, you're going to be involved in something big. Almost a decade ago the damages caused by expansive soils in the United States exceeded $10 billion. The Chinese estimate damages caused to their railroad system alone are 100 million yuan (Chinese currency) annually. But efforts to control these damages with geomembranes have been tried with success, and you are not alone. The documentation of over 100 U.S. projects can aid you in your work.

## 16.1  A Partial List of Dos

A partial list of dos might start and finish with test, test, test! That said, the first item would be to check the site on the ground. Be fully aware of what is planned for the development. Check the relation between existing natural ground and the proposed elevations of what's to be built. The on-site walking inspection should involve careful observation for signs or symptoms of expansive soil damage on nearby structures or substantial cracking in the existing ground. Are building walls plumb or tilting? Are spaces opening up between one part of the structure and another? If it's a transportation facility, are there signs of repeated patching, cracking, or waves in the pavement? Is the base showing up where there was pavement, or is the railroad ballast revealing intrusion of the swelling soil subgrade? These are indications that further investigation is required.

A search of data available from a variety of agencies and societies is another key item on the do list. The USDA Soil Conservation Service provides valuable information on substantial earth data. Universities and state and federal highway and airport agencies have funded studies and valuable reports that provide vital information on the presence of expansive soils. One major study funded by the Federal Highway Administration conducted by the U.S. Army Corps of Engineers, Waterways Experiment Station, provides information covering the entire United States. The National Academy of Science's Transportation Research Board is another valuable source of information, as is the Federal Housing Administration. The American Society of Civil Engineers, Technical Activities Committees, with several institutes and the International Society of Soil Mechanics and Foundation Engineers, have much supporting information. Worldwide engineering groups are examining the expansive soil challenge too, and there are frequent international and regional geotechnical gatherings.

A good local library with a proper electronic search system can provide a series of sources of useful information. Professional engineering societies further increase the scope of the information search. None of this is wasted time or money. This body of existing research can prepare the engineer, designer, and the entire design and construct team for building and maintaining facilities in the vicinity of expansive soils.

It is also important to secure results from adequate test holes on the site. This could be as simple as making one test hole on a lot for a single-family home. Or it could be as involved as making several test holes for a larger, more expensive home or a multifamily development or larger commercial or public building. All test holes should be at least 20 feet below the bottom finished subgrade grade of the structure.

For highways, a test hole should be made every quarter of a mile. Laboratory tests should follow standard ASTM methods or those of the appropriate governing agency. Atterberg limits are probably the initial and most common initial tests run. Their results can quickly provide guidelines as to what might be expected. Subgrade strength compressibility and other laboratory tests as appropriate should be conducted and assessed. Suction tests are also now gaining increased attention as informative tests.

When it is decided to use a geomembrane, tests for the material should be run. These tests may frequently be an ASTM series for tensile strength, elongation at failure, puncture strength, permeability coefficient, trapezoid tear, and thickness. Their numbers in similar order are ASTM D 4995, 4833, 4491, 4533, and 1777. Some states such as Texas have developed their own specifications, and they are sometimes used by other states. In all cases the designer and constructor should be sure that the specification is current and applicable.

A reminder at this point is valuable: To the test, test, test caveat should be added document, document, document. This could be photographs from the field inspection phase or photocopies of newspaper reports on cases in the area involving legal claims stemming from damages caused by expansive soils. The laboratory test values should be documented along with an assessment of the results. If this sounds a bit paranoiac, so be it. Better this, though, than to be a defendant in a lawsuit.

Another point to keep in mind should you decide to use a geomembrane is to be sure to extend it for a long enough stretch—that is, make the length sufficient to avoid moisture intrusion around the end of the fabric. The same precaution applies to placing the geomembrane deep enough. Verify the depth of the zone of activity, of significant moisture change, at the site. Also, be certain when dealing with buildings that the location of service pipes is known, and prepare to create moisture barriers around these potential water sources. In addition, maintain a thorough construction inspection routine. The best plans in the world need to be followed up by good construction inspection. The horror book has many tales of no reinforcement in the slab or of reinforcement on the subgrade rather than at a specified distance from the subgrade or of posttensioned steel with no tension at all. Be aware of the questions raised about geomembranes controlling the expansive soils. The concern about water building up underneath the geomembrane and damaging the pavement is a valid one. Note, however, that not one of the 100 projects case histories has reported this as a problem. Another remaining question involves whether the backfill of the trench will become a destructive water source again, leading to more pavement damage. Again, not one of the 100 project engineers has reported this to be a problem.

## 16.2 A Partial List of Definitely Don'ts

A partial list of definitely don'ts is worthy of substantial thought. Avoid assumptions not backed by tests of the proposed site. Just because the lot next door hasn't moved or has not had laboratory tests indicating expansive soils, don't presume the site you're planning to improve has a similar soil condition. It may have. Then again, it might not. Definitely don't make an assumption on the situation in order to save a test or save some time. Costs of failure to identify an expansive soil condition can reach very high dollar amounts. Class-action suits related to damages caused by expansive soils have been filed in areas from the Texas Gulf coast to the front range in Colorado and to California from Contra Costa County to the Imperial Valley just east over the coastal range from San Diego. If you're an engineer and believe that additional testing is required, don't be put off by the words "we can't afford it." As the world-renowned geotechnical engineer Fu Hua Chen has said, "Don't take that client."

Try to avoid thinking "We've always done it that way" in coming to a design decision. It's quite possible that the distortion to a structure, a building, or a roadway has no relation to an expansive soil. The subgrade might indeed be rock or a nonexpansive soil. The existence of a certain type of expansive soil a mile down the road or even on the adjacent lot does not guarantee that the soil is the same where you're working. There should be a concern, an awareness, and an unwillingness to save a penny when the end might be spending a lot more. Furthermore, do not avoid reading current literature on the subject.

Someday there might be a wonderful patented fluid to take the expansive qualities away from the clays and shales exhibiting these destructive forces. Until then, be careful when presented with miracle solutions. Don't fail to keep your wits, professional reputation, and money about you.

## 16.3  Further Dos and Don'ts

A full synopsis of what needs to be done to avoid falling in to class-action suits and other liability problems is a challenge to express in 25 words or fewer. The first recommendation is to review this chapter's list of dos. The next thought would be to read the next heading in this chapter, the don'ts. One leading geotechnical engineer told the following story. He had been subjected to several suits regarding his designs in some expansive soils areas. He transferred all his assets to his wife except the calculators in his office. He then dropped his professional liability insurance. When the next complainant came to his office, he responded by saying that he had no insurance company liability and he had no possessions except the computers in the office. Claims and suits disappeared. Many geotechnical engineers have not chosen to follow such a lead. Class-action suits continue. Increasingly claims filed by the damaged parties against the lawyers who represented them for taking too large a cut of the award have grown.

It has been suggested that a stipulation be added to any design contract, a series of steps that would be required to have been taken before the path would lead to the engineer who designed the geotechnical provisions. Such steps would include maintaining adequate drainage away from the facility and keeping vegetation distances away from the facility. These measures would vary depending on the particular type of planning. These are just some of the relevant matters that could be added to the contract.

## 16.4 Conclusion

These dos and don'ts are excellent advice for all those involved with expansive soils including the engineers, builders, owners, agencies, insurance companies, and the taxpayers who ultimately pay the bill for all public facilities. Expansive soils are a huge and growing global problem. In this century increased attention has been focused toward recognition, theoretical understanding, and solutions addressing the challenge. Avoiding the areas, replacing the soil, treating it, and isolating it from moisture changes seen as the root of the volumetric changes are all available options. Studies, articles, reports, and books have been written on these questions. The challenge remains to pick the right solution for the particular problem at hand.

Avoiding the expansive soils is frequently not a real-world solution. The other options have all met with success and failure. The use of geomembranes to control damages caused by expansive soils has increased and is generally viewed as successful. When it hasn't seemed to work, the reasons appear to be clear. The greatest need for the decision-making process is adequate testing. There's no point in using a method to control something that isn't there. Also, remember over 100 projects in the United States have used geomembranes to control expansive soils. Most have been on highway projects, but others have involved buildings, railroads, and even an airport here and there. If all the signs are right, wind up, and try a geomembrane—you'll like it.

# Appendix A
# Vertical Moisture Barrier

The following is excerpted from the Texas Department of Transportation, Departmental Materials Specification D-9-6210.

## 1. Description

This Item shall govern for the materials, composition, quality, sampling, and testing of vertical moisture barrier as specified herein. The moisture barrier shall consist of geomembrane in the form of sheeting or geotextile in the form of coated fabric or that of a fabric-sheeting laminate.

## 2. Materials

1. *General.* The moisture barrier shall be constructed exclusively of man-made materials. When sheeting is furnished, it shall be of single-layered construction, without seams, and shall be formulated of man-made materials complying with the detailed specifications set forth. Fabric where furnished shall be made of either woven or nonwoven thermoplastic fibers. Such fabrics shall be furnished precoated on one or both sides or impregnated so as to make the fabric impermeable to water or moisture under the conditions of test set forth in this speci-

fication. Fabric-sheeting laminate being furnished shall consist of fabric fused or heat-scaled to sheeting so as to form an integral geomembrane. The moisture barrier must be able to withstand normal handling and placement at material temperatures from −7 to 63°C (20–145°F) without endangering the serviceability of the material in the intended application. If the moisture barrier evidences delamination such delamination may serve as grounds for rejection. The moisture barrier furnished shall be mildew, abrasion and puncture resistant and suitable for long term burial in the presence of water and/or moisture in the intended construction application.

## 3. Physical Requirements

The moisture barrier furnished shall meet the following additional requirements when sampled and tested in accordance with the methods specified.

| | Original Physical Properties | Test Method | Requirement |
|---|---|---|---|
| 1. | Fabric Weight, on an ambient temperature air-dried tension-free sample, expressed as kg/m$^2$ (oz/sq yd) | Texas Test Method Tex-616-J. "Testing of Construction Fabrics" | 0.22 minimum (6.5 minimum) |
| 2. | Water permeability, as determined when moisture barrier is subjected to the equivalent of a 3.0 m (10 ft) column of water for a period of two (2) hours. Moisture passing through the fabric is determined by weight gain of desiccant. Permeability is expressed as weight of in kg/m$^2$ (oz/sq yd) of vertical moisture barrier. | Tex-616-J | 0.02 maximum (0.6 maximum) |
| 3. | Abrasion resistance. After prescribed sandblast, the moisture barrier shall meet the specification requirement for water permeability. Expressed as weight of water in kg/m$^2$ (oz/sq yd) of vertical moisture barrier. | Texas Test Method Tex-851-B, "Method for Evaluating the Abrasion Resistance of Pavement Marking Materials" modified as follows: 152 mm (6 in) sample distance, 40 psig regulated blast pressure, and one kilogram of blast medium with a blast time of two (2) minutes plus or minus 15 seconds per one kilogram of blast medium. | 0.02 maximum (0.6 maximum) |

| | Original Physical Properties | Test Method | Requirement |
|---|---|---|---|
| 4. | Load characteristics at break or at 100% elongation, whichever occurs first under the conditions of test. Material shall meet specified minimum in both machine direction (MD) or cross-machine direction (CMD), test values to be expressed in Newton (pounds). | ASTM D 1682, Grab Test G with 25 mm×51 mm (1 in×2 in) jaws and constant time to break rate of extension of 20 ± 3 seconds, as specified. | 667 minimum (150 minimum) |
| 5. | Apparent elongation at break or rupture, expressed in %. | ASTM D 1682, Grab Test G with 25 mm×51 mm (1 in×2 in) jaws and constant time to break rate of extension of 20 ± 3 seconds, as specified. | 20% minimum |
| 6. | Tear strength, determined by the tongue (single rip) method on specimens prepared from "as-received" samples. Specimens are to be tested at a cross-head speed of 305 ± 12 mm/minute (12 ± 0.5 in/minute). Test results are to be calculated by the "average of five (5) highest peaks" method. *Both* the average of five (5) specimens cut with the longer dimension parallel to the machine direction (MD) *and* the average of five (5) specimens cut in the cross-machine direction (CMD) shall meet the specified minimum expressed in Newton (pounds). | ASTM D 751 | 15 minimum |

## 4. Packaging Requirements

The moisture barrier shall be packaged in rolls of the length and width specified on the plans, as directed by the Engineer or in the purchase order awarded by the State. The material furnished on a given roll shall be one piece construction. Individual pieces of moisture barrier on a given roll being joined together by splicing, lapping, bonding, stapling, etc., will not be acceptable. The moisture barrier itself shall be uniformly wound onto suitable cylindrical forms or cores to aid in handling and unrolling. Each roll of fabric and the form or core upon which it is rolled shall be packaged individually in a suitable sheath, wrapper or container to help protect the fabric from damage due to ultraviolet light and moisture during normal storage and handling.

## 5. Labeling or Tagging

Each roll shall be identified by a tag or label securely affixed to the outside of the roll on one end. This tag or label must list the following required information:

1. A unique roll number, serially designated
2. Manufacturer's lot number or control numbers, if any
3. Name of fabric manufacturer
4. Date of manufacture
5. Brand name of the product
6. Manufacturer's style or catalog designation of the fabric, if any
7. Roll width in millimeters or meters (inches)
8. Roll length in meters (yards)
9. Gross weight of entire package which is to include moisture barrier, core, wrapping and sheath or container identification tag, etc.
10. Tare weight of core, wrapping, sheath or container identification tag, etc.
11. Net weight of fabric alone

Example:

| | | |
|---|---|---|
| **(a)** Roll No. 31275 | **(e)** "Fabriweld" | **(i)** Gross 66.7 kg (147 lbs) |
| **(b)** Lot 290 control 6740 | **(f)** "300-X" | **(j)** Tare 8.2 kg (18 lbs) |
| **(c)** Afgan Fabrics | **(g)** Width 3.8 m (150 in) | **(k)** Net Wt. 58.5 kg (129 lbs) |
| **(d)** Jan. 16, 1987 | **(h)** Length 91.4 m (100 yd) | |

## 6. Sampling and Testing Requirements

Samples for testing purposes shall be taken in accordance with Test Method Tex-735-1. Testing shall be in accordance with Test Method Tex-616-J or the test methods stated herein.

The values stated in either SI units or English units are to be regarded as standard. Within the text, the English units are shown in parentheses. The values stated in each system are not exact equivalents; therefore, each system shall be used independently of the other.

## 7. Basis for Rejection

Should any individual sample selected at random from 100 rolls, or fraction thereof, fail to meet any specification requirement, then that roll shall be rejected and two additional samples shall be taken, one from each of two other additional rolls selected at random from the same 100-roll lot, or fraction thereof. If either of these two additional samples fail to comply with any portion of the specification, then the entire quantity of rolls represented by that sample shall be rejected.

## 8. Measurement and Payment

1. *Procurement by the state.* Measurement and payment for all materials under this specification shall be in accordance with the conditions prescribed in the purchase order awarded by the State.
2. *Contracts.* Measurement and payment of all materials governed by this specification and utilized in the performance of work specified in the contract shall be paid for in accordance with the pertinent specification, vertical moisture barrier, in the contract.

# Appendix B
# Impermeable Plastic Membrane

The following is excerpted from the Wyoming Department of Transportation, Standard Specifications for Road and Bridge Construction, 1996 edition, pages 542 through 631. It was adopted by the Transportation Commission of Wyoming on June 15, 1995.

## Section 630: Impermeable Plastic Membrane

**630.01 Description.** This work shall consist of furnishing and placing impermeable plastic membrane in accordance with these specifications and as shown on the plans, or as specified by the Engineer.

**630.02 Materials.** The membrane shall be woven or nonwoven polypropylene or polyethylene geotextile with a bonded polypropylene or polyethylene film and shall conform to the requirements of subsection 712.23.

If a sand cushion is required, it shall conform to the material requirements shown on the plans. The moisture content of the sand shall be optimal for compaction.

**630.03 Construction Requirements.** The foundation for the membrane shall be smooth and free of pockets, loose rocks, or any other material which could damage the membrane.

Longitudinal and lateral joints shall be overlapped a minimum of 150 mm and glued with a product that conforms to the membrane manufacturer's recommendations for product type, application rate and curing procedures. If glued joints are not utilized, the membrane shall be overlapped a minimum of 0.6 m. The overlaps shall be shingled so that the exposed edges face in the same direction as the flow of drainage. All blemishes, holes or scars shall be reinforced and sealed with waterproof plastic adhesive tape. All unacceptable portions of the membrane and all unacceptable sealed joints shall be replaced without additional payment.

Pegs or pins shall not be used to hold the membrane in place. Membrane which is damaged following installation as a result of the construction operations shall be replaced without additional payment.

During placement of the membrane, equipment may run directly on the membrane, however, no turning movements will be allowed. The membrane shall not be placed during rain or wet weather conditions or when the subgrade is wet.

When sharp or angular rock or aggregate will be in contact with the membrane, a sand cushion shall be provided between the membrane and the subgrade as called for on the plans. The sand cushion shall be a minimum of 100 mm in thickness.

For vertical installation in trenches, the backfill material shall come from the trench excavation and shall be placed in a manner to prevent damage to the membrane. Backfill materials in direct contact with the membrane shall be free of large rocks, and shall be approved by the Engineer prior to backfilling. The trench shall be backfilled as soon as the membrane is in place. The backfill material shall be compacted and finished in accordance with subsection 203.035.

The cover material shall be placed over the membrane within five days.

During periods of shipment and storage, the membrane rolls shall be enclosed in heavy-duty wrapping to protect the membrane from direct sunlight, ultraviolet rays, temperatures greater than 60°C, mud, dirt, dust and debris. Any membrane left unprotected shall be removed from the project.

The product name, type of material and the lot or batch identification shall be clearly labeled on each roll.

Test results and a manufacturer's certification showing the membrane performance in regard to the material requirements of this specification shall be submitted to the Engineer and Geology Program. At least two weeks before the use of any membrane, a sample of the membrane 2 m in length by the full width of the roll shall be submitted to the Geology Program through the Engineer. The sample shall be labeled with product name, machine direction, the lot and batch number, date of sampling, project number, and certification of compliance with the material specifications.

**630.04 Method of Measurement.** Impermeable plastic membrane will be measured in place by the square meter of surface area, including vertical surfaces in trenches, with no allowance for overlaps. The trench excavation and backfill, and the sand cushion will be subsidiary to the pay item.

**630.05 Basis of Payment.** Payment will be made under:

| Pay item | Pay unit |
|---|---|
| Impermeable plastic membrane | $m^2$ |

## Section 631: Slotted Drains

**631.01 Description.** This work shall consist of the construction of slotted drains fabricated from corrugated, galvanized-steel pipe in accordance with these specifications and as shown on the plans.

**631.02 Materials.** Materials shall conform to the following subsections:

| | |
|---|---|
| Corrugated Steel Pipe and Pipe Arches | 707.02 |
| Structural Steel | 713.01 |

**631.03 Construction Requirements**

**631.031 Fabrication.** The slotted drains shall be fabricated from galvanized, 2-mm-thick steel pipe, as specified. The grate assembly shall be fabricated from structural carbon steel in conformance with Section 501. Metal end caps and coupling bands shall be galvanized steel of the same thickness as the pipe.

Concrete end plugs and concrete used as backfill around the slotted drain may be of any of the classes of structural concrete used on other portions of the project or it may be local ready-mixed concrete containing not less than 300 kg/$m^3$ of portland cement meeting the requirements of Class C concrete. Any concrete used in conjunction with the installation of the slotted drains will be subsidiary to the pay item.

A close-riveted and soldered annular pipe or continuously-welded helical pipe shall be used for the slotted drain. The pipe coupling bands and end plugs shall be watertight. Gaskets shall be used with the coupling bands to insure water tightness. A metal or concrete end plug shall be used. Either type of end plug shall match the corrugations of the pipe and be watertight.

**631.032 Installation.** The slotted drains shall be placed in the locations and at the elevations specified on the plans. The pipe shall be supported to allow free flow of the concrete around the pipe. The concrete shall be placed in two lifts. The first lift shall be to the spring line of the pipe. When this lift has set sufficiently to prevent floating or movement of the pipe, the remainder of the concrete shall be placed. The two-step operation may be eliminated if leveling devices are used that are capable of holding the pipe in place.

Excavation and backfill shall be in accordance with Section 206.

When placing the curb or paving material over the slotted drain, the opening shall be covered to prevent foreign material from entering the slot. The slot shall not extend above the paving material or curb and gutter section.

**631.04 Method of Measurement.** Slotted drains will be measured by the meter of pipe. The concrete joint seal material and other material will be subsidiary to the pay item.

| Item | Size | Material type | Surface finish |
|---|---|---|---|
| U Bolt Clamps | | Commercial | |
| Hardware | | Commercial | ASTM A 153 275 |
| Single, double, multiple supports | 50 mm OD 14 gage | ASTM A 513 | ASTM A 525 M 275 |
| Socket | 12 gage | ASTM A 569 M | Dip-coated with rust-inhibiting primer Fed Spec TT-P-636 |
| Wedge | 12 gage | ASTM A 513 | ASTM A 525M 275 |
| Mounting bracket and angle | 14 gage | ASTM A 366M | ASTM A 525M 275 |
| Adapter plate | 12 gage | ASTM A 366M | ASTM A 525M 275 |
| Anti-twist plate | 16 gage | ASTM A 366M | |
| Cantilever pipe, galvanized. Pressure testing not required. | | ASTM A 53 Type F, Schedule 40 | ASTM A 525M 275 |
| Screw fittings may be steel or malleable iron. Pressure testing not required. | | ASTM A 858M or ASTM A 47M Grade 22010 | ASTM A 153 Class C or D |

**712.22 Vacant**

**712.23 Geotextile and Impermeable Plastic Membrane.** Geotextile and impermeable plastic membrane shall meet the following requirements as applicable for the specified use.

## Geotextile and Impermeable Plastic Membrane Specifications
### Geotextile and Membrane Requirements (Minimum Roll Averages)

| Geotextile and membrane property | Test method | Drainage & filtration | Erosion control | Silt fence | Separation & stabilization | | Embankment & retaining wall reinforcement | Impermeable plastic membrane |
|---|---|---|---|---|---|---|---|---|
| | | | | | Woven | Non-woven | | |
| *Performance Criteria During Service Life* | | | | | | | | |
| Equivalent or Apparent Opening Size, US Standard Sieve, μm | ASTM D 4751 | 425–150 | 425–106 | 850–300 | 425–150 | 425–150 | 425–150 | — |
| Thickness, μm | ASTM D 5199 | — | — | — | — | — | — | 225 |
| Permittivity, $s^{-1}$ | ASTM D 4491 | 0.2 | 0.1 | 0.01 | 0.01 | 0.1 | 0.005 | $<10^{-6}$ mm/s[1] |
| Retention Efficiency, % | ASTM D5141-91 | — | — | 75 | — | — | — | — |
| *ASTM D 5141-91 Strength Requirements* | | | | | | | | |
| Wide-Width Strip Tensile Strength, kN/m | ASTM D 4595[b] ASTM D 4885[g] | 7 | 11 | 11 | 22 | 11 | 35 | 14[9] |
| Elongation at Failure, % | ASTM D 4595[b] | 40 min. | 40 min. | 20 min. | 20 min. | 20 min. | 35 max. | 20 min. |
| Burst Strength, kPa | ASTM D 3786 (Diaphragm Method) | 900 | 1450 | 1725 | 2000 | 1450 | 2960 | — |
| Trapezoid Tear Strength, N | ASTM D 4533 (Any Direction) | 110 | 180 | 220 | 220 | 180 | 330 | 220 |
| Puncture Strength, N | ASTM D 4833[c] | 110 | 220 | 220 | 330 | 220 | 490 | 265 |
| Seam Efficiency, % | ASTM D 4595[b] ASTM D 4884 | 100 | 100 | 100 | 100 | 100 | 100 | — |

Geotextile and Impermeable Plastic Membrane Specifications
Geotextile and Membrane Requirements (Minimum Roll Averages)

| Geotextile and membrane property | Test method | Drainage & filtration | Erosion control | Silt fence | Separation & stabilization | | Embankment & retaining wall reinforcement | Impermeable plastic membrane |
|---|---|---|---|---|---|---|---|---|
| | | | | | Woven | Non-woven | | |
| | | Environmental Requirements | | | | | | |
| Mildew/Rot Resistance, % | AATCC 30 1988[o] | 100 | 100 | 100 | 100 | 100 | 100 | 100 |
| Insect/Rodent Resistance, % | AATCC 24 1985[o] | 100 | 100 | 100 | 100 | 100 | 100 | 100 |
| Ultraviolet Resistance, % Strength Retention | ASTM D 4355 | [d] | 90 | 70 | [d] | [d] | [d] | —[d] |

# Appendix C
# Special Provision for Impermeable Plastic Membrane

The following is excerpted from the Wyoming Department of Transportation, Special Provision for Impermeable Plastic Membrane, Project No. 033-3(7), Cody-Montana State Line, Pat O'Hara Creek, Park County.

This Special Provision supplements, amends and where in conflict therewith, supersedes Section 629—GEOTEXTILE FABRIC of the Standard Specifications.

## Description

This work consists of placing an impermeable plastic membrane at the locations and to the dimensions shown on the plans, in accordance with this Special Provision.

## Materials

The membrane will be woven or non-woven polypropylene or polyethylene fabric with a bonded polypropylene or polyethylene film, and will have the following typical minimum properties:

| | | |
|---|---|---|
| Wide Width Tensile Strength | 80 lbs/in | ASTM D-4595 |
| Elongation at Failure | >20% | ASTM D-4595 |
| Puncture Strength | 60 lbs | ASTM D-4833 |
| Permeability Coefficient | $<10^{-7}$ cm/sec | ASTM D-4491 |
| Trapezoid Tear | 50 lbs | ASTM D-4533 |
| Thickness | 12 mils | ASTM D-1777 |

A sample of the impermeable plastic membrane will be submitted to the Engineer in accordance with the requirements of Subsection 629.03—CONSTRUCTION of the Standard Specifications.

## Construction

The foundation for the impermeable plastic membrane will be smooth and free of pockets, loose rocks, or any other material which could damage the fabric. The surface shall be free of any excess moisture that may allow slippage of the membrane.

All longitudinal and lateral joints will be overlapped a minimum of six inches (150 mm) and glued with a product in accordance with the manufacturer's recommendations for product type, application rate and curing procedure or overlapped a minimum of two feet (0.61 m). The overlaps will be shingled so that the exposed edges face in the same direction as the flow of drainage. All blemishes, holes or scars will be reinforced and sealed with waterproof plastic adhesive tape. All unacceptable portions of the impermeable plastic membrane, and all unacceptable sealed joints will be replaced at the Contractor's expense.

Impermeable plastic membrane which is damaged following installation as a result of the Contractor's operations will be rejected, and will be replaced at the Contractor's expense.

Trench backfill material will come from the trench excavation, and will be placed in a manner that will prevent damage to the impermeable plastic membrane. Backfill material in direct contact with the membrane will be free of large rocks and will be approved by the Engineer prior to backfilling. The trench will be backfilled as soon as the impermeable plastic membrane is in place. The backfill material will be compacted and finished in accordance with Subsection 203.10—CONSTRUCTION OF EMBANKMENT AND TREATMENT OF CUT AREAS WITH MOISTURE AND DENSITY CONTROL of the Standard Specifications.

Special Provision for Impermeable Plastic Membrane

## Method of Measurement

Impermeable plastic membrane will be measured in place by the square yard of surface area, including vertical trench walls, with no allowance for overlaps.

## Basis of Payment

The accepted quantities of impermeable plastic membrane will be paid for at the contract unit price per square yard, complete, in place; which payment will be full compensation for excavating and backfilling the vertical trenches, furnishing and placing the impermeable plastic membrane, and all labor, equipment, tools and incidentals necessary to complete the work.

Payment will be made under:

| Pay item | Pay unit |
|---|---|
| Impermeable Plastic Membrane | SY |

Geotextile (Fabric) and Impermeable Plastic Membrane Specifications
Fabric and Membrane Requirements (Minimum Roll Averages)

| Fabric and membrane property | Test method | Drainage & filtration | Erosion control | Silt fence | Separation & Stabilization | | Embankment & retaining wall reinforcement | Impermeable plastic membrane |
|---|---|---|---|---|---|---|---|---|
| | | | | | Woven | Non-woven | | |
| | | Performance Criteria during Service Life | | | | | | |
| Equivalent or Apparent Opening Size, US Standard Sieve (μm) | ASTM D 4751 | 40–100 (425–150) | 40–140 (425–106) | 20–50 (850–300) | 40–100 (425–150) | 40–100 (425–150) | 40–100 (425–150) | — |
| Thickness, Mils (μm) | ASTM D 1777 | — | — | — | — | — | — | 12 (305) |
| Permittivity, Sec$^{-1}$ | ASTM D 4491 | 0.2 | 0.1 | 0.01 | 0.01 | 0.1 | 0.005 | $<10^{-7}$ cm/sec[f] |
| Retention Efficiency, % | VTM-51-79[a] | — | — | 75 | — | — | — | — |
| | | Strength Requirements | | | | | | |
| Wide Width Strip Tensile Strength, lbf./in. (kN/m) | ASTM D 4595[b] ASTM D 4885[g] | 40 (7) | 65 (11) | 65 (11) | 130 (22) | 65 (11) | 200 (35) | 80[g] (14) |
| Elongation at Failure, % | ASTM D 4595[b] | 40 min. | 40 min. | 20 min. | 20 min. | 20 min. | 35 max. | 20 min. |
| Burst Strength, psi (kPa) | ASTM D 3786 (Diaphragm Method) | 130 (900) | 210 (1450) | 250 (1725) | 290 (2000) | 210 (1450) | 430 (2960) | — |
| Trapezoid Tear Strength, lbf. (N) | ASTM D 4533 (Any Direction) | 25 (110) | 40 (180) | 50 (220) | 50 (220) | 40 (180) | 75 (330) | 50 (220) |
| Puncture Strength, lbf. (N) | ASTM D 4833[c] | 25 (110) | 50 (220) | 50 (220) | 75 (330) | 50 (220) | 110 (490) | 60 (265) |
| Seam Strength, lbf./in (kN/m) | ASTM D 4884 | 40 (7) | 65 (11) | 65 (11) | 130 (22) | 65 (11) | 200 (35) | — |

Geotextile (Fabric) and Impermeable Plastic Membrane Specifications
Fabric and Membrane Requirements (Minimum Roll Averages)

| Fabric and membrane property | Test method | Drainage & filtration | Erosion control | Silt fence | Separation & Stabilization | | Embankment & retaining wall reinforcement | Impermeable plastic membrane |
|---|---|---|---|---|---|---|---|---|
| | | | | | Woven | Non-woven | | |
| | | Environmental Requirements | | | | | | |
| Mildew/Rot Resistance, % | AATCC 30 1988[e] | 100 | 100 | 100 | 100 | 100 | 100 | 100 |
| Insect/Rodent Resistance, % | AATCC 24 1985[e] | 100 | 100 | 100 | 100 | 100 | 100 | 100 |
| Ultraviolet Resistance, % Strength Retention | ASTM D 4355 | [d] | 90 | 70 | [d] | [d] | [d] | [d] |

[a]Virginia Department of Highways and Transportation test method.
[b]8″ (200 mm) wide×4″ (100 mm) length specimen tested at a strain rate of 10% (0.4 in.) (10 mm) per minute.
[c]Using 5/16″ (8 mm ± 0.01 mm) diameter flat tipped steel cylinder centered with ring clamp.
[d]Non-stabilized or low susceptible geotextiles should not be exposed to ultraviolet radiation for more than 5 days.
[e]American Association of Textile Chemists and Colorists test procedures.
[f]Permeability Coefficient (ASTM D 4491).
[g]Wide Strip Tensile Method for Geomembranes.
NOTE: Values are omitted where the requirement does not apply to the end use application or is sufficiently addressed by a related property.

# Appendix D
# Geomembrane

The following is excerpted from the Arizona Department of Transportation, geomembrane specifications.

## Description

The work under this item includes furnishing labor, equipment, and materials to construct a moisture barrier as detailed in the project plans and these special provisions. It also includes trench excavation, placement of the membrane, backfill and compaction. The purpose of the moisture barrier installation will be to restrict water infiltration from the roadway ditches and roadway prism to the underlying soils.

## Material Requirements

The geomembrane moisture barrier shall be supplied in accordance with and conform to the general requirements listed in Section 1014-1 of the Standard Specifications.

The moisture barrier shall consist of an impervious sheet of fabric-reinforced geomembrane, of composite construction, with polyethylene film or sheet bonded, between two woven or nonwoven polypropylene or polyester fabrics.

The geomembrane shall be inert to chemicals and hydrocarbons and shall be resistant to mildew, rot, ultra violet exposure, insects and rodents. It shall also conform to the minimum average roll values for properties listed in the following table:

| Property | Requirements | Test method |
| --- | --- | --- |
| Width, ft. | 9 | N.A. |
| Thickness, mils | 14 | ASTM D1777 |
| Grab Tensile Strength, lbs., at break or 100% elongation, whichever occurs first | 170 | ASTM D4632 |
| Grab Elongation at break, % | 20 | ASTM D4632 |
| Puncture Strength, lbs. | 70 | ASTM D4833 |
| Burst Strength, psi | 250 | ASTM D3786 |
| Trapezoidal Tear, lbs. | 40 | ASTM D4533 |
| Permittivity | 0 max. | ASTM D4491 |

Minimum average roll values represent the average test results for a lot in the weaker direction when sampled according to ASTM D4354 and tested according to the test method specified above.

Samples of the geomembrane shall be submitted for testing. No samples shall be taken within 5 feet of either end of a roll. Samples shall be a minimum of 3 feet long by the full roll width. A minimum of one sample shall be taken per lot. More samples may be required as determined by the Engineer.

The contractor shall submit a sample of the proposed geomembrane, selected on the basis of material property requirements previously listed, prior to use. A maximum of three geomembrane products will be evaluated and tested by ADOT. Each submittal shall include product information sheets and Certificate of Analysis as required by 1014-1 of the Standard Specifications.

## Construction Requirements

The geomembrane shall be installed in accordance with the plans, these special provisions, and the manufacturer's installation procedures and recommendations.

## Weather Limitations

Geomembrane shall not be placed when weather conditions, in the opinion of the Engineer, are not suitable to allow placement or installation. This will normally be at times of wet or snowy conditions, heavy rainfall, extreme cold or frost conditions, or extreme heat.

Geomembrane

## Equipment

Mechanical or manual laydown equipment shall be capable of handling full rolls of fabric, and laying the fabric smoothly, without wrinkles and folds, in the position specified. The equipment shall be in accordance with the manufacturer's recommendations or as approved by the Engineer.

## Subgrade Preparation

The subgrade shall be prepared in accordance with Section 203-3.03(A) of the Standard Specifications, prior to placement of the geomembrane.

## Geomembrane Placement

The installation of the geomembrane shall be in accordance with the following sequence (Figs. D-1 through D-3):

1. In areas requiring geomembrane placement in a vertical trench, trenches will first be excavated to the minimum dimensions listed in the plans details. Geomembrane material shall be placed against the inside wall of the excavated trench, extending the full depth of the trench and a minimum of 24 inches onto the horizontal subgrade surface. Geomembrane which extends onto the horizontal subgrade surface shall be fixed to the subgrade surface with stakes, nails or

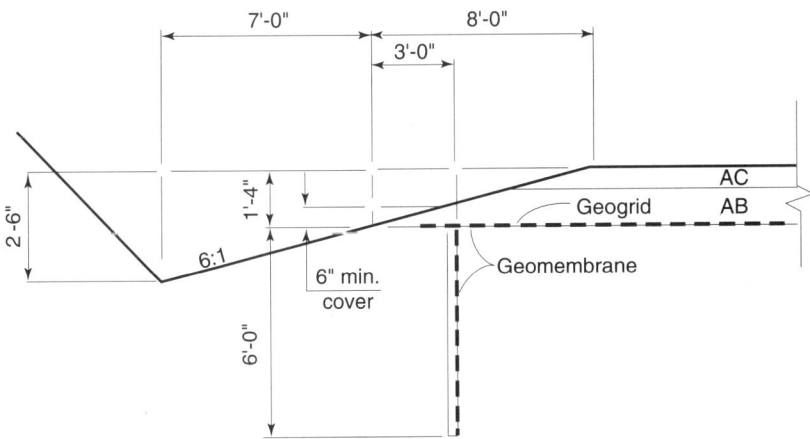

**Figure D-1.** Typical section, Claypool, Jakes Corner Highway, 188 G1 261 H2157SID (PLH-038-1)(17)(C).

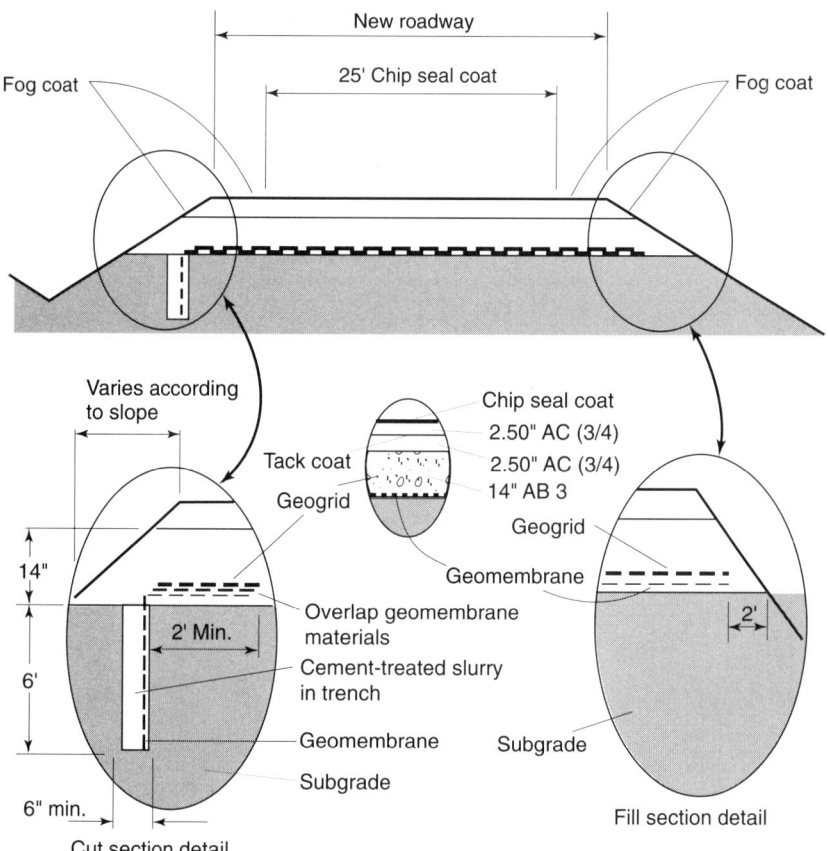

**Figure D-2.**

other method approved by the Engineer to secure the geomembrane in place during backfilling of the trench. Immediately following the geomembrane placement, the vertical trench shall be backfilled with cement-treated slurry conforming to Section 501-3.02 of the Standard Specifications to within one foot of subgrade elevation. No backfilling above the cement-treated slurry shall be commenced until 24 hours after its placement. The final one foot of the trench shall be backfilled with excavated material, after it is processed to ± 2% of optimum moisture content. This backfill shall be compacted to 95% of the maximum density determined in accordance with the requirements of Arizona Test Methods 225, 226, 227, 230 or

Geomembrane

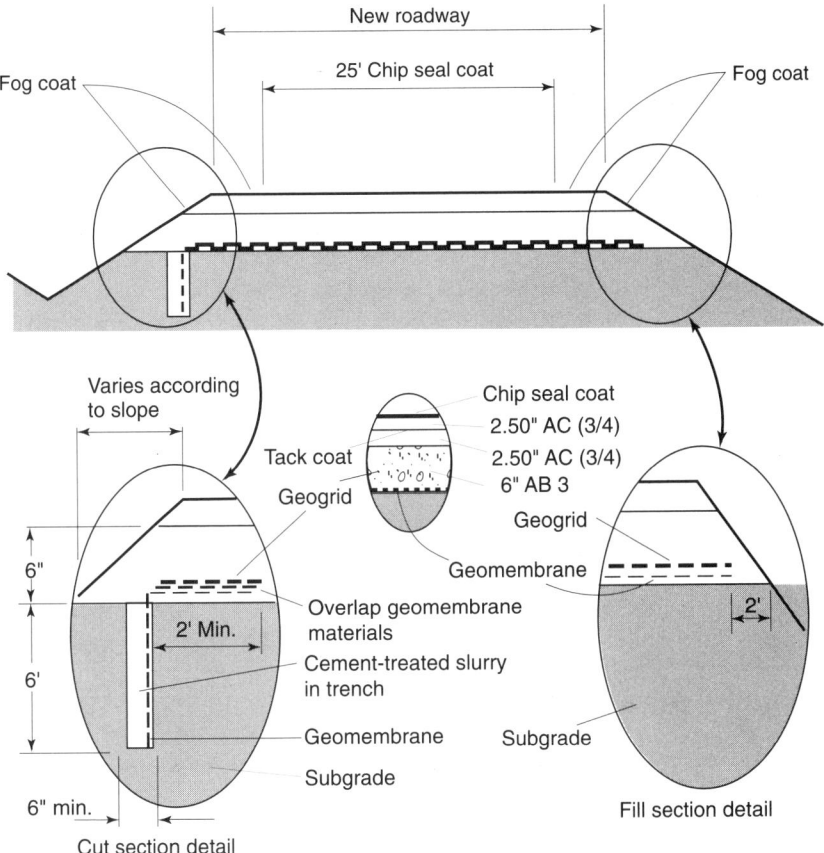

**Figure D-3.**

231, and 232. Extreme care shall be taken in avoiding damage to the geomembrane supported on the trench wall. Removal and replacement of any geomembrane that is damaged will be the responsibility of the contractor.

2. Geomembrane shall then be placed over the horizontal subgrade surface to the extent shown on the plans details. All wrinkles and folds shall be removed and overlaps shall be minimum 24 inches wide. Any geomembrane which extends onto the subgrade from a vertical trench shall also be overlapped a minimum of 24 inches. Sewn seams will not be permitted unless it can be demonstrated that they are watertight factory seams.

## Placement and Compaction of Aggregate

Aggregate materials shall be placed by back dumping the aggregate in a manner which does not damage the geomembrane. The aggregate material shall be spread in a minimum thickness of 8 inches onto the geomembrane in a constant forward direction. Traffic or construction equipment shall not be permitted directly on the geomembrane unless approved by the Engineer for emergency purposes. Pins or piles of aggregate can be used to hold the geomembrane in place while being covered.

Overstressing the subgrade soil shall be avoided by utilizing equipment in spreading and dumping that exerts only moderate pressures on the soil. If ruts of 2 inches or greater occur in the aggregate, the contractor shall use lighter equipment which transmit less ground pressure. Any ruts which develop during spreading or compacting aggregate shall be filled with additional aggregate rather than bladed from adjacent areas so that the final design aggregate thickness is maintained. Construction equipment shall not be allowed to turn or stop suddenly on the aggregate placed over the geomembrane. Aggregate base shall be compacted to a density not less than 95% of the maximum density. Aggregate base material shall not be mixed or processed on the geomembrane. The aggregate base material shall be premixed at the stockpile area or at another location in a manner approved by the Engineer. Aggregate base materials will be sampled for acceptance after premixing and prior to placement of the geomembrane. Contamination and segregation of aggregate base materials prior to or during placement shall be minimized.

Any damage to the geomembrane occurring during placement of the aggregate must be repaired immediately. The aggregate shall be removed from the damaged area to allow placement of a geomembrane patch extending 3 feet on all sides beyond the damaged area, followed by replacement of the aggregate.

## Method of Measurement

The geomembrane shall be measured for payment by the square yard, complete in place. No additional measurement or allowance will be made for material in overlaps or seams.

## Basis of Payment

The accepted quantities of geomembrane, measured as provided above, will be paid for at the contract unit price, which shall include full compensation for furnishing all labor, equipment, and materials involved in placement of the geomembrane as shown in the project plans. No measurement or payment will be made for geomembrane that has been damaged, and/or contaminated.

# Appendix E
# Geotextile Fabric

The following is excerpted from the Mississippi State Highway Department, Special Provision 907-714-32, Code IS, December 1, 1989.

Section 714, Miscellaneous Materials, of the Standard Specifications is amended as follows:

Delete Subsection 714.13, page 859, and substitute:

## 907-714.13 Geotextile Fabrics

### 907-714.13.1 General

Unless specified otherwise, the fabric may be woven or nonwoven. The fabric shall consist only of long chain polymeric yarns or filaments such as polypropylene, polyethylene, polyester, polyamide or polyvinylidenechloride and shall be formed into a stable network such that the yarns or filaments retain their relative position. The fabric shall be mildew resistant and inert to biological degradation and naturally encountered chemicals, alkalies and acids. Fabric which is not protected from sunlight after installation shall contain stabilizers and/or inhibitors to make it resistant to deterioration from direct sunlight, ultraviolet rays and heat.

The edges of the fabric shall be selvaged or finished in such a manner to prevent the outer yarn or filaments from raveling. The fabric shall be free of defects or flaws which affect the required physical properties.

Fabric for silt fence shall be manufactured in widths of not less than three feet and fabric for other applications shall be manufactured in widths of not less than six feet. Sheets of fabric may be sewn or bonded together at the factory or other approved locations but deviation from the physical requirements will not be permitted.

Tests for manufacturer's certification shall be conducted with fabric as shipped by the manufacturer and acceptance testing will be conducted with fabric from the project.

### 907-714.13.2 Geotextile Fabric for Silt Fence

The fabric shall conform to the physical requirements of Type I or II as shown in Table I. Unless a specific type is specified in the plans or contract documents, the Contractor may select Type I or II.

**907-714.13.2.1 Woven Wire Backing.** Except as provided herein, silt fence shall be reinforced with a woven wire backing. The wire backing shall be at least 32 inches high and have no less than six horizontal wires. Vertical wires shall be spaced no more than 12 inches apart. The top and bottom wire shall be 10 gage or larger. All other wire shall be no smaller than 12 1/2 gage.

### 907-714.13.9 Shipment and Storage

During shipment and storage, the fabric shall be protected from direct sunlight, ultraviolet rays, temperatures greater than 140°F, mud, dirt, dust and debris. The fabric shall be wrapped and maintained in a heavy-duty protective covering.

### 907-714.13.10 Manufacturer's Certification

The Contractor shall furnish to the Engineer three copies of the manufacturer's certified test reports and certification that each lot in a shipment complies with the requirements of the contract. All fabric, steel pins, washers, fence posts, woven wire and wire staples are subject to approval by the Engineer upon delivery to the work site and prior to incorporating in the work.

### 907-714.13.11 Acceptance Sampling and Testing

Final acceptance of each shipment will be based on results of tests performed by the Department on verification samples submitted from the

Geotextile Fabric

project. The Engineer shall select one roll at random from each shipment for sampling. A sample extending full width of the randomly selected roll and containing at least five square yards of fabric shall be obtained and submitted by the Engineer. The sample from each shipment shall be provided at no cost to the State.

Section 907-486, Pavement Fabric, is added to the 1976 Standard Specifications for Road and Bridge Construction as follows:

## 907-486
## Pavement Fabric

### 907-486.01 Description

This work shall consist of furnishing and installing a geotextile fabric in accordance with details shown on the plans and requirements of the contract.

### 907-486.02 Material

The fabric for this work shall meet the requirements of Subsection 907-714.14.

### 907-486.03 Equipment

The Contractor shall provide equipment necessary for placing the fabric on a smooth subgrade and in the position and location set out in the plans.

### 907-486.04 Construction Details

The area shall be prepared to establish a relatively smooth surface. If required by the Engineer sand may be placed over these areas to cushion the fabric. The fabric shall be placed as smooth as possible. Wrinkles and folds in the fabric shall be removed by stretching and staking as required.

The strips of the fabric shall be overlapped a minimum of 18 inches for each joint. Securing pins with washers shall be inserted through both strips of overlapped cloth along a line through the mid-point of the overlap at intervals required by the Engineer to prevent movement of the fabric until covered.

The subsequent courses of material shall be back-dumped in such a manner as to avoid damage to the underlying fabric.

## 907-486.05 Method of Measurement

The accepted fabric placed in accordance with these specifications and as directed will be measured by the square yard. Laps will not be measured for payment.

## 907-486.06 Basis of Payment

The fabric will be paid for at the contract unit price per square yard. This price shall be full compensation for furnishing and placing the fabric, pins, lapping and maintaining the fabric until covered, and satisfactorily completing the work specified.

Payment will be made under: Pay Item No. 907-485-C: Geotextile Fabric for Moisture Barrier—per sq. yd.

Section 714, Miscellaneous Materials, of the Standard Specifications is amended as follows:

Add Subsection 907-714.14

## 907-714.14 Geotextile Fabric for Moisture Barrier

### 907-714.14.1 General

The geotextile fabric for moisture barrier shall consist of sheeting, coated fabric or a fabric sheeting laminate constructed exclusively of man-made materials. Sheeting shall be of single-layered construction. Coated fabric shall be made of woven or non-woven polyester or polypropylene. The fabric shall be furnished precoated on one or both sides or impregnated so as to make the fabric impermeable to water or moisture. Fabric-sheeting laminate shall consist of fabric fused or heat-sealed to sheeting so as to form an integral geotextile membrane.

The geotextile fabric shall be able to withstand normal handling and placement at material temperatures from 20°F to 145°F without endangering the serviceability of the material in the intended application. If the geotextile evidences de-lamination, such de-lamination may serve as grounds for rejection. The geotextile fabric shall be mildew, abrasion, and puncture resistant and suitable for long term burial in the presence of water and/or moisture in the intended construction application. It shall be packaged in rolls of the length and width specified on the plans or directed by the Engineer.

Geotextile Fabric

## 907-714.14.2 Physical Requirements

The geotextile fabric shall meet the following additional requirements when sampled and tested in accordance with the methods specified.

| Original physical properties | Method | Requirements |
|---|---|---|
| Fabric weight, oz/sq.yd. (air-dried tension-free sample) | Texas test method Method Tex-616-J "Testing construction fabrics" | 6.5 minimum |
| Water permeability expressed as weight of water in oz/sq.yd. Fabric is subjected to the equivalent of a ten foot column of water for a period of two hours. Moisture passing through the fabric is determined by weight gain of disiccant. | Tex-616-J | 0.6 maximum |
| Abrasion resistance expressed as weight of water in oz/sq.yd. After prescribed sandblast the fabric shall meet the requirement for water permeability. | Texas test method Tex-851-B "Method for Evaluating the Abrasion Resistance of Pavement Marking Materials" modified as follows: six-inch sample distance, 40 psig regulated blast pressure and one kilogram of blast medium with a blast time of two minutes plus or minus 15 seconds per one kilogram of blast medium. | 0.6 maximum |
| Load characteristics at break or 100% elongation, whichever occurs first. Material shall meet specified minimum in both machine direction and cross-machine direction. Test values to be expressed in pounds. | ASTM D 1682, Grab Test G with 1'×2' jaws and constant time to break rate of extension of 20 plus or minus three seconds, as specified. | 150 minimum |

| | Text | |
|---|---|---|
| Original physical properties | Method | Requirement |
| Apparent elongation at break or rupture, expressed in percent | See above Grab Test G | 20% minimum |
| Tear strength determined by the tongue (single rip) method on specimens prepared from "as-received" samples. Specimens are to be tested at a cross-head speed of twelve plus or minus 0.5 inches/minute. Test results are to be calculated by the "average of five highest peaks" method. Both the average of five specimens cut with the longer dimension parallel to the machine direction and the average of five specimens cut in the cross-machine direction shall meet the specified minimum expressed in pounds. | ASTM D 751 | 15 minimum |

## 907-714.14.3 Packaging Requirements

The geotextile fabric shall be packaged in rolls of the length and width specified on the plans or directed by the Engineer. The material shall be uniformly wound onto suitable cylindrical forms or cores to aid in handling and unrolling. Each roll shall be packaged individually in a suitable sheath, wrapper or container to protect from ultraviolet light and moisture damage during normal storage and handling.

## 907-714.14.4 Identification

Each roll of fabric or container shall be visibly labeled with the name of the manufacturer, type of geomembrane or trade name, date, lot number and length, width and quantity of material.

Geotextile Fabric

### 907-714.14.5 Sampling

A sample of five square yards of the fabric shall be furnished to the State from each shipment for verification testing. The samples shall be provided at no cost to the State.

### 907-714.14.6 Certification

The Contractor shall furnish to the Engineer three copies of the manufacturer's certified test report(s) showing results of all required tests and certification that the material meets the specifications. Certification shall be furnished for each lot in a shipment.

Section 714, Miscellaneous Materials, of the Standard Specifications is amended as follows:

Type II fabric may be installed without the wire backing provided:

A. Post spacing reduced to six feet or less.
B. The fabric manufacturer recommend its use without the wire backing.
C. The fence posts are inclined toward the run-off source but not more than 20° from vertical.
D. The fabric shall be attached to the posts as recommended by the manufacturer.

### 907-714-13.2.2 Posts

Wood or steel posts may be used. Wood posts shall have a minimum diameter of three inches and length of five feet and shall be straight enough to provide a fence without noticeable misalignment. Steel tee posts shall be five feet long, approximate 1 3/8 inches wide, 1 3/8 inches deep and 1/8 inch thick with a nominal weight of 1.33 pounds per foot prior to fabrication. The posts shall have projections, notches, or holes for fastening the wire backing or fabric to the posts.

### 907-714-13.2.3 Staples

Staples shall be made of nine gage wire with a minimum length of one inch after bending.

### 907-714-13.3 Geotextile Fabric for Subsurface Drainage

Unless otherwise specified, the fabric shall conform to the physical requirements of Type III as shown in Table I.

### 907-714-13.4 Geotextile Fabric Undersell

The fabric shall be nonwoven polyester or polypropylene which is satisfactory for use with asphalt cements. Unless otherwise specified, the fabric shall conform to the physical requirements of Type IV in Table I.

### 907-714-13.5 Geotextile Fabric for Use under Riprap

Unless otherwise specified, the fabric shall conform to the physical requirements of Type V in Table I. The requirements for tensile, bursting, puncture and trapezoidal tear strengths may be reduced 50 percent when the fabric is cushioned from rock placement by a 6 inch minimum layer of sand.

### 907-714-13.6 Geotextile Fabric Stabilization

The fabric shall meet the physical requirements as shown in TABLE I for the type specified in the plans or contract documents.

### 907-714-13.7 Securing Pins

Steel pins used for anchoring the fabric shall be three-sixteenth inch in diameter, minimum length of 15 inches, pointed at one end and fabricated with a head for retaining a steel washer. A minimum one and one-half inch washer shall be installed on each pin.

### 907-714-13.8 Identification

Each roll of fabric or container shall be visibly labeled with the name of the manufacturer, type of fabric or trade name, lot number and quantity of material.

### 907-714-13.9 Shipment and Storage

During shipment and storage, the fabric shall be protected from direct sunlight, ultraviolet rays, temperatures greater than 140°F, mud, dirt, dust and debris. The fabric shall be wrapped and maintained in a heavy-duty protective covering.

### 907-714-13.10 Manufacturer's Certification

The contractor shall furnish to the Engineer three copies of the manufacturer's certified test reports and certification that each lot in a shipment complies with the requirements of the contract. All fabric, steel pins, washers, fence posts, woven wire and wire staples are subject to approval by the Engineer upon delivery to the work site and prior to incorporating in the work.

### 907-714-13.11 Acceptance Sampling and Testing

Final acceptance of each shipment will be based on results of tests performed by the Department on verification samples submitted from the project. The Engineer shall select one roll at random from each shipment for sampling. A sample extending full width of the randomly selected roll and containing at least five square yards of fabric shall be obtained and submitted by the Engineer. The sample from each shipment shall be provided at no cost to the State.

**Table E-1.**

| Physical properties | Type designation | | | | | | | Test method |
|---|---|---|---|---|---|---|---|---|
| | I | II | III | IV | V | VI | VII | |
| Thickness, mils | — | — | — | 40 | — | — | — | ASTM D 1777 |
| Weight, oz./sq. yd. | — | — | — | 4-9 | — | — | — | ASTM D 3776, Option A or B |
| Asphalt retention, oz./sq. ft. | — | — | — | 3.0 | — | — | — | Miss. Test Method MT 64 |
| Maximum change in area, percent | — | — | — | 15 | — | — | — | Miss. Test Method MT 64 |
| Permeability, cm/sec (See Note 2) | — | — | 0.01 | — | 0.01 | 0.01 | 0.01 | AASHTO M 288 (Appendix) |
| Flow rate, gal./min./sq. ft. (See Note 2) | — | — | 30 | — | 30 | 30 | 30 | AASHTO M 288 (Appendix) |
| Equivalent opening size (EOS) (See Notes 2 and 3) | | | | | | | | Miss. Test Method: MT 60 |
| Woven fabric | 20-100 | 20-100 | 40-100 | — | 70-100 | 70-100 | 70-100 | |
| Nonwoven fabric | 20+ | 20+ | 40+ | — | 70+ | 70+ | 70+ | |
| Tensile strength after ultraviolet exposure, lbs | 40 | 80 | — | — | — | — | — | ASTM D 4632 (CRE) after 500 hours exposure on xenon arc weatherometer as detailed in ASTM G 26 (See Note 1) |

*Note 1:* A test result shall be the average of the test values of five specimens.
*Note 2:* Unless designated otherwise in the plans or contract documents.
*Note 3:* The EOS test for nonwoven fabric may be waived by the Testing Engineer.
*Note 4:* All of the above strength tests except "retained strength" are to be conducted in a dry condition.

**Table E-2.** Geotextile Fabrics Minimum Average Roll Value

| Physical properties | Type designation | | | | | | | Test Method |
|---|---|---|---|---|---|---|---|---|
| | I | II | III | IV | V | VI | VII | |
| Tensile strength, lbs. (weaker principal direction) | 50 | 90 | 90 | 90 | 200 | 280 | 450 | ASTM D 4632 (CRC) (See Note 1) |
| Elongation at required strength, percent. | — | 50 (Max.) | — | 20 | — | 50 | — | ASTM D 4632 (CRC) (See Note 1) |
| Bursting strength, psi | 100 | 180 | 140 | — | 300 | 450 | 700 | ASTM D 3786 Diaphragm Bursting Tester |
| Puncture strength, psi | — | — | 35 | — | 80 | 110 | 180 | ASTM D 3787, Tension Test-in. Machine with Ring Clamp; Steel Ball replaced with a 5/16 inch hemispherical tip |
| Trapezoidal tear, lbs | — | — | 35 | — | 65 | 100 | 150 | ASTM D 4533 (CRE) (See Note 1) |
| Retained strength when wet, percent. | 100 | 100 | 100 | — | 100 | 100 | 100 | ASTM D 4632 (CRE) and ASTM D 3786 and 3787, as above. (See Note 1) |

# Index

Airfields, affected by expansive soils, 44, 141–142
Aitchison, Gordon, 38, 74
Amarillo, Texas, 63, 73–75
American Society of Civil Engineers, 24, 28, 49, 60, 69, 71, 91, 172
American Society of Testing Materials, 9, 53, 62, 64
Amoco Fabrics, 12–14, 158
Argentina, expansive soils in, 20
Arizona Department of Transportation, 4, 32, 53, 62, 126–127, 150, 195
Aronowitz, M., 2
Asphalt applications, 5, 61, 93–94, 116, 150
Atterberg, A., 28, 47
Atterberg tests, 23, 27–28, 47, 51–52, 56, 94, 96, 98, 102
Auger drills, 50–51
Austin, Texas, 63, 75
Australia:
　expansive soils in, 2–3, 19–21, 25, 36, 38, 42, 65, 69, 74, 89, 91, 130–131
　geomembrane use in, 5–6, 12–14, 25, 38, 63, 76–83, 130–137, 162
Australian methods, costs of, 156, 158–160
Australian Road Research Board, 38, 135, 137

Backfill, 64, 82–84, 115, 151–154
Backhoes, 151
Barenberg, Ernest, 63
Bathurst, Richard, 38, 74
Beams, foundations for, 71
Bedrock, 72–73, 88, 90
Belgium, expansive soils in, 5
Bell footings, 88
Bentonite, 17, 72–73, 109, 124–125
Berger, William, 56
Bexar County, Texas, 112, 152

BIDIM, 12
Bitumin geomembranes, 137
Black, 35
Black clay, 2
Black cotton soil, 2
Blight, Geoffrey, 25–26, 39, 74
Bonaparte, Rudolph, 15, 169
Bonded fibers, 10, 12
Brakey, B. A., 5, 62, 94
Brazil, expansive soils in, 3, 20, 24
Bridges, affected by expansive soils, 88
Bryant, Frank, 57
Buchanan, Spencer, 36
Buckingham, 35
Building codes, 49
Building Research Advisory Board, 56, 58, 71, 75, 87, 91
Buildings:
　with geomembranes, 69–84
　without geomembranes, 85–91
Burma, expansive soils in, 2–4, 19–21, 25, 56, 70, 91

California, canals in, 144
California Department of Transportation, 32, 127–128
Canada:
　expansive soils in, 2–3, 19–21, 37–38, 69, 74, 91
　geomembrane use in, 14–15
Canals, affected by expansive soils, 44–45, 142–144
Cantilever beam design approach, 56
Capillary potential, 35
Casagrande, Leo, 4, 49, 150
Case trencher, 159
Cation-exchange capacity, 2–3, 18, 27
Cement applications, 61, 150
Center for Highway Research, 29

**215**

Checklists, 171–175
Chemical barriers, 78–83
Chen, Fu Hua, 2–3, 18, 24–25, 27, 35, 48–49, 52, 57, 61, 71, 87, 89, 162, 174
China:
  expansive soils in, 1–2, 20–21, 25, 42, 44, 61, 171
  geomembrane use in, 6, 13–14, 144–145
Chinle clay, 90
Chlorinated polyethylene geomembranes, 9
Chlorosulfonated polyethylene geomembranes, 9
Civil engineering, 4, 28, 47
Class-action suits, 2, 16, 21, 24, 42, 48, 85–87, 89, 174
Clays, 2, 4, 17–18, 27–28, 34–35, 61, 90
Claystone, 2
Clotella Canal, 144
Cody shale, 122
Collingwood, Victoria, 78–79
Colombia, expansive soils in, 3, 24, 44, 75, 141–142
Colonial Hills development, 42
Colorado Department of Transportation, 5–6, 13, 32, 43, 62, 93–95, 150
Colorado Geological Survey, 49, 71–73
Comer, Alice (coauthor on U.S. Bureau of Land Reclaimation), 13, 15, 25, 45, 49, 72, 142–144, 169
Commercial buildings (see Institutional buildings)
Construction, 52–53
Contra Costa County, California, 16, 24, 86–87, 89
Corpus Christi, Texas, 2, 16, 24, 42, 44–45, 87, 89, 146
Costs, 64, 139, 153, 155
  accuracy of projections, 157–158
  of Australian methods, 156, 158–160
  of damage, 1, 22–23, 41, 171
  factors affecting, 155
  placement rates, 160
  of Texas methods, 156–160
  treatment methods, 161–163
Cottonwood Dam, 144
Count, 10
Crack surveys, 130–131, 135
Croney, 35
Culberson County, Texas, 109–110

Dallas, Texas, 44, 113

Dallas-Fort Worth Intercontinental Airport, 44, 61, 142, 150
Dane Chemco, 13
De Wet, 39
DeBerardino, Stephen (coauthor with Hawkins, W.M.), 15, 169
Deep vertical fabric moisture barriers, 6, 89, 102–113, 116, 118, 124, 139, 152, 156, 160, 163, 168
Degradation, 165
Delaware, geomembrane use in, 5, 15
DeMarco, Sandra, 136
Dempsey, Barry, 63
Denver, Colorado, 16, 21, 24, 42, 47, 73, 86–87, 89
Denver International Airport, 43–44, 142, 150
Design, 52–53
  applicability/feasibility of approaches, 64–66
  expansive soil control options, 58–62
  general control considerations, 55–58
  geomembrane placement, 62–64
Diapers, 6
Differential scanning calirometry, 167
Dikes, 6, 14, 45, 60, 143
Dipping bedrock, 73, 88, 90
Ditch Witch trencher, 124, 129, 136, 151–152, 159
Donaldson, G. W., 24
Dougherty, John, 116
Drainage, 59, 66
Drilled-shaft foundations, 43, 56–57, 88–89
Drilling rigs, 149
Drilling for test sites, 50–51
Du Pont Company, 5–6, 12–13, 43, 53, 62, 94, 97, 99, 102, 105, 109, 145, 150, 156, 166
Duderstadt, Ed, 56, 71, 88
Duderstadt Drilling Company, 43
Durability, 14–15, 165–169
Dynaflect testing, 101

Eagle Ford shale, 113
Electroosmosis, 5, 61–62, 150
Encapsulation, 5
Engineered fibers, 10
England, expansive soils in, 3
Environmental issues, 21–22
Ethiopia, expansive soils in, 2, 25
Ethyl vinyl acetate, 11–12, 97
Ethylene interpolymer alloy geomembranes, 9

Index

Evans, Robert, 38, 135–136, 153, 159
Expansive shales, 17
Expansive Soil Technical Advisory Group, 24, 33, 94–95
Expansive soils:
  damage caused by, 21
  and design stage, 55–66
  disasters, 23
  environmental impact, 21–22
  facilities affected by, 41–45
  financial impact, 22–23, 41
  geographical distribution, 1–3, 18–20
  and highways, 43–44, 60, 93–118, 121–139
  and institutional buildings, 55–56, 87–89
  physical properties, 17–18
  problem solving attempts, 4–5
  and residences, 55–56, 85–87
  studies of, 3–4, 23–40
  terminology, 17–18
  testing, 47–54
Expert system, 59
Exxon Chemical Company, 166

Fabrene geomembranes, 13, 123
Facilities affected by expansive soils:
  airfields, 44, 141–142
  bridges, 88
  canals, 44–45, 142–144
  highways, 43–44, 60, 93–118, 121–139
  institutional buildings, 56–57, 87–89
  pipelines, 44–45, 145–146
  railroads, 44–45, 144–145
  residences, 55–56, 85–87
  types of, 41–43
Falcon Homes, 86
Falk, M. A., 15
Federal Aviation Agency, 44, 141
Federal Highway Administration, 17, 19, 24, 32–33, 43, 49, 94, 165, 172
Federal Housing Administration, 49, 56, 71, 172
Fibers, 10–11
Filaments, 11–12
Financial issues, 22–23
(*See also* Costs)
Flooding of proposed sites, 90–91
Florida Department of Transportation, 32
Flowable backfill, 82–84
Fly ash applications, 61, 108–109, 150
Focht, John, 29
Formed fibers, 10

Fort Collins, Colorado, 72
Fort Union shale, 121–122
Fort Worth, Texas, 44
Foundations:
  drilled-shaft, 43, 56–57, 88–89
  pier, 55–56
  steel-reinforced, 90
France:
  expansive soils in, 3, 45
  geomembrane use in, 6, 13, 145
Fraser, 58
Fredlund, Delwyn, 37, 74
Friant Kern Canal, 45, 144
Frydman, S., 39, 75

General McMullen Drive, 62–63, 97–102, 151, 156–158
Geomechanics Seminar on Road Engineering, 134–135
Geomembranes:
  applications of technology, 12–13
  buildings with, 69–84
  buildings without, 85–91
  capabilities/limitations, 14–15
  costs of, 155–163
  design stage, 55–66
  development of technology, 5–6, 11–12
  durability of, 165–169
  effectiveness, 15–16
  fabrics/manufacturers, 13–14
  facilities used in, 41–45, 141–146
  geographical distribution of use, 14
  highways with, 93–118, 121–139
  new developments, 83–84
  placement of, 5–6, 62–64, 149–154
  remediation, 75–83
  selection of, 53–54
  specifications, 195–201
  terminology, 9–11
Geoseal geomembranes, 111
Geosynthetics Conferences, 54
GeoTech Institute, 49
Geotextile fabrics, 203–213
Gerhardt, Manfred, 71
Germany, geomembrane use in, 13, 22
Gibsite, 18
Gilgai, 65
Gillette, H. S., 28–29
Giroud, J. P., 10
Gradall backhoe, 107
Grading, 66

Greenville, Texas, 108–109
Grim, 18
Guadalupe County, Texas, 95–96, 111–112, 152
Gulf Building, 71
Gypsum board, 56

Hager, G. M., 15
Halliburton, 165
Handy, Richard, 74
Hart, S. S., 60, 71
Hawkins, W. M. (coauthor with Stephen DeBarardino), 15, 169
Hayden Rhodes Aqueduct, 143
Herbicides, 81
Higgins, C. M., 61
High-density polyethylene geomembranes, 10, 80, 83, 135, 143, 169
Highlands Ranch development, 86–87
Highways, affected by expansive soils, 43–44, 60, 93–118, 121–139
Holden, James, 38, 74, 77–79, 83, 131, 135, 137, 153
Holland, J. E., 74, 131
Hollandrain trencher, 159
Holtz, W. G., 22, 24–25, 60, 71–72
Hoop, Guy, 6
Horizontal placement, 5–6, 62–64, 149, 151–154
Houston, Texas, 5, 150
Hudspeth County, Texas, 109, 150
Hungary, expansive soils in, 20

ICI, 12
Illite, 17
Imperial Valley (California), 16, 24
Impermeability, 10–11, 14
Impermeable plastic geomembranes, 183–193
India, expansive soils in, 2, 19–20, 70, 75
Industrial Fabrics Association, 54
Institute of Civil Engineers (South Africa), 3
Institute of Technology (Israel), 39, 74
Institutional buildings, 56–57, 87–89
International Conference on Expansive Soils, 19, 22, 25, 36–37, 41, 70, 145
International roughness index, 65
International Society of Soil Mechanics and Foundation Engineers, 4, 28, 49–50, 70, 172

Israel:
  expansive soils in, 2–3, 17, 19, 21, 25, 36, 39, 42, 44–45, 69, 74–75, 141–142, 145
  geomembrane use in, 5, 12–14, 39, 130–137

Jackson, Mississippi, 149
Jayatilaka, R., 135
Jeff Davis County, Texas, 109–110
Jennings, Jeremiah, 36, 39, 74
John Deere backhoe, 102
Johnson, Lawrence, 33, 49
Jones, D. E., 22, 24–25
Jones, Jesse, 71
Jordan, expansive soils in, 2, 42

Kaolinite, 17–18
Kassiff, G., 18, 25, 34–35, 39, 49, 52, 131
Kelly, Conard, 4, 60, 93–94, 161
Kentucky Department of Transportation, 32, 129
Kitti, 20
Koerner, Robert, 9
Komornik, Amos, 39, 49, 75
Krazinski, 26
Krishnen, 2

Laboratory testing, 59
Lambe, T. W., 28, 49
Landfills, 21–22
Lawsuits (*see* Class-action suits)
Leakage, 11
Leonards, 38, 74
Lime applications, 4–5, 60–61, 93–95, 97, 108–109, 116, 149–150, 161–162
Lime slurry pressure injection, 61
Limestone scalpings, 111
Linear density, 10
Liquifill, 82–83
Little, Dallas, 61, 150, 161–162
Livneh, Moshe, 18, 25, 34–35, 39, 44, 49, 75
Lousiania Department of Transportation, 61, 109
Love Canal (New York), 22
Low-density polyethylene geomembranes, 10, 80
Lytton, Robert, 29, 37, 56, 58, 73
Lytton procedure, 58

# Index

MacDonald, Thomas, 28, 32–33
Maintainers, 158
Mangos shale, 5
Mann, Anthony, 38, 65
Marienfeld, Mark (authored geomembrane report), 5–6, 12–16, 32, 75–76, 93–118, 121–139, 162–163
Mass per unit area, 10
Massachusetts Institute of Technology, 28
McAllister Freeway, 106
McDonald, Eugene, 32, 44, 61, 63, 95, 130
McDowell, Chester, 29, 31, 56, 61, 96–97
McKeen, R. G., 36, 44, 141–142
McManus, Kerry, 38, 65, 83, 153
Melbourne, Victoria, 81–83
Melbourne International Airport, 136
Mexico, expansive soils in, 3, 19–20, 75, 144
Meyer, Kirby, 60, 63, 75–76, 89, 162
Michelin Tire Company, 116
Midwest Canvas, 13, 129
Miller, Debora Hamburg, 17, 52, 57–58, 61, 162
Mirafi geomembranes, 12–13, 106, 108, 113–114, 123, 127, 129
Mission Viejo development, 42, 47, 86–87
Mississippi Department of Transportation, 5, 32, 53, 60, 128, 149–150, 203
Mitchell, F. W., 74
Moisture measurement devices, 114–115
Moncos shale, 43
Monomers, 11
Montana Department of Transportation, 32, 127
Montmorillonite, 2, 17–18
Morgenstern, 74
Morocco, expansive soils in, 24
Morton Plains (Victoria), 130–131
Motor vehicle test tracks, 116
Mudstone, 2
Mulch gardens, 66

Nachlinger, D. R., 29
National Academy of Sciences, 172
National Historic Civil Engineering Award, 71
National Lime Association, 94
National Science Foundation, 71
Navajo Reservation, 90, 127
Navarro County, Texas, 95
Nebraska, canals in, 144
Needle-punched fabrics, 11–12
Nelson, John, 17, 49, 51–52, 57–58, 61, 162

Netherlands, expansive soils in, 6, 12, 14
Nevada Department of Transportation, 128–129
Nevels, James, 32, 36
New Mexico Department of Transportation, 32, 128
Nicolon/Mirafi, 13
Nisichi, Eng, 75
Nix Medical Office and Hospital Building, 57
Noble Park, Victoria, 79–80
Noe, D. C., 88
Nonwoven fabrics, 11–12
North American Geosynthetics Society, 54
North Las Vegas, Nevada, 44, 142
Novastrene, 13

Ofer, Z., 51, 74
Oklahoma Department of Transportation, 32, 60, 161
Olefins, 11
Oman, expansive soils in, 19, 42, 75
Open-pit holes, 51

Parsons trencher, 105
Particle size, 18
Patrick, David, 33
Peck, R. B., 34, 39
Penetration tests, 51
Perry trencher, 159
Peru, expansive soils in, 3, 20, 75
Petromat geomembranes, 12–14, 109–111, 128, 158
Phillips, 12–14, 109–111, 128, 158
Phipps, John, 56
Phoenix, Arizona, 42, 90, 143
Photologging, 98, 100–101, 103
Pierre shale, 95, 122, 129
Piers, foundations for, 55–56, 71
Pipelines, affected by expansive soils, 44–45, 145–146
Placement of geomembranes, 5–6, 62–64, 149–154
Placement production rates, 160
Plasticity indices, 4–5, 27, 30, 47, 51–52, 94, 96–97, 102, 131
Plasticizers, 11
Poland, expansive soils in, 20, 75
Polyethylene geomembranes, 10, 75, 80, 83, 135
Polyflex, 13

Polymerization, 11
Polymers, 11
Polypropylene geomembranes, 10, 12, 79, 97–98
Polyurethane, 81
Polyvinyl chloride, 11, 79–80, 143
Polyvinyl chloride geomembranes, 10, 12–13, 15
Ponding, 5, 60, 66, 96–97, 149
Port Philip, Victoria, 80–81, 83
Porter, Henry, 4–5, 21, 28–29, 31, 36, 47, 49, 60, 95–96
Post Tensioning Institute, 56, 58, 73, 75, 87, 91
Post Tensioning Institute procedure, 58–59
Postconstruction testing, 97, 101, 103, 105, 108–109, 111, 118
Potassium applications, 62
Potential vertical rise tests, 23–24, 56
Power spectral density, 38, 65
Preconstruction testing, 97, 101, 110
Profilometers, 108–109, 135
Psychrometers, 109

Qatar, expansive soils in, 19, 75

Railroads, affected by expansive soils, 44–45, 144–145
Ranen, 52
Recycled materials, 13, 22
REEMAY, 12–13, 112, 166
Reese, Lyman, 88
Regar, 2
Rehabilitation contracts, 105–113, 116, 123, 127–129
Remediation, 75–83, 131
Residences, 55–56, 85–87
Rhone Poulence, 12
Ribbons, 11
Richmond Homes, 86–87
Road graders, 158
Roads (see Highways)
Root inhibitors, 78–83, 91, 133–135
Rumania, expansive soils in, 20
Russia:
    expansive soils in, 20, 45, 75
    geomembrane use in, 6, 145

St. Kilda, Victoria, 80–81, 83

San Antonio, Texas, 2, 16, 42, 44–45, 57, 60, 62–63, 66, 87–89, 97–108, 112–113, 156, 168
San Antonio International Airport, 44
Sands, water-bearing, 116
Saudi Arabia:
    expansive soils in, 1–2, 19, 42, 75, 87
    geomembrane use in, 6
Scottsdale, Arizona, 143
Scrim, 10
Seed, B., 49
Serviceability indices, 98–99, 101, 103–108, 116
Shales, 4–5, 17–18, 43, 73, 95, 113, 121–122, 129
Shelby tubes, 51
Shrinkage index, 52
Shrinking soils, 17
Siemienko, Roger, 6
Simons, N. E., 39
Simpson, Willard, 4, 21, 36, 42–43, 56–57, 69–71, 87–88, 162
Singh, 2
Site classification, 59
Slab design, 57–58
Slabs, floor, 90
Slim-line trenchers, 38, 81, 83–84, 136–137, 153–154, 163
Smectite, 17
Smith Young Tower, 42–43, 57, 88
Snethen, Donald, 33, 49
Soil classification systems, 18
Soil Conservation Service, 49, 172
Soil inspection, on-site, 50
Soil mechanics, 4, 28, 47
South Africa, expansive soils in, 2–3, 19, 21, 25–26, 31, 36, 39, 42, 45, 69–70, 74, 90–91
South Australia, highways in, 136–137, 153
South Dakota Department of Transportation, 5, 32, 44, 61, 93–95, 129–130
Spain, expansive soils in, 3
Spreadability indices, 98–99, 101
Spun-bonded fibers, 10, 12, 97–98
Stabilizers, 11
Steele shale, 122
Steenburgen trencher, 159
Streets (see Highways)
Structural engineering, 52–53
Subgrades, destructive, 64–65, 93–95
Suction tests, 23, 34–36, 51, 135

Sudan, expansive soils in, 2, 19, 25–26, 42
Sunraysia Highway, 131–135
Survivability ratings, 165
Swelling clays, 4, 17
Swinburne procedure, 58, 91
Swinburne University of Technology, 137, 139

Tan, Harry, 12
Tar paper, 96
Technion, 59, 74
Teeth, trencher, 136–137, 159
Tel Aviv, Israel, 139
Teng, Paul, 31, 60, 97, 166
TERAM, 12
Terzaghi, Karl, 4, 27–28, 33–34, 43, 47, 49, 70–71
Terzaghi Dam, 15
Testing:
 in Australia, 168–169
 and design/construction phases, 52–53
 evaluation of results, 50–52
 and geomembrane selection, 53–54
 importance of, 47–48
 information on, 48–50
 by manufacturers, 166–167
 methods, 50–52
 number of sites, 50–52
 on-site soil inspection, 50
 postconstruction, 97, 101, 103, 105, 108–109, 111, 118
 preconstruction, 97, 101, 110
 by state agencies, 167–168
Texas A & M University, 4, 28, 32, 36–37
Texas Department of Transportation, 4–6, 13, 21, 31–32, 53, 61–63, 93–118, 150–154, 156–159, 161, 163, 168–169, 177
Texas Geological Survey, 49
Texas methods, costs of, 156–160
Texas Transportation Institute, 118
Thompson, R., 36
Thornthwaite moisture index, 23, 38, 65
Tisinger, L. G., 15
Townsend, Frank, 33
Transit Tower (*see* Smith Young Tower)
Transportation Research Board, 34, 48
Tree roots, problems caused by, 78–83, 91, 133–135
Trench walls, sliding of, 108, 114, 137
Trenchers, 151–153
Trifluration, 81

Tucson, Arizona, 42, 90, 144
Tullamarine, Victoria, 80, 135–136
Tungsten-coated chisel teeth, 136–137, 159
Typar geomembranes, 5, 10, 12–13, 15, 43, 53, 62, 94–95, 97–102, 105, 110, 112, 122–123, 127–128, 145, 150, 156, 166–167

Underdrains, 56, 64, 75
Uniform Building Code, 58
Uniroyal Company, 116
United States:
 expansive soils in, 1–4, 19–22, 24–25, 29, 31–33, 36–37, 42–45, 69, 71–75, 85–91, 142–146
 geomembrane use in, 5–6, 12–16, 32, 75–76, 93–118, 121–139, 162–163
U.S. Army Corps of Engineers, 2, 24, 33, 36, 43, 49, 162, 172
U.S. Bureau of Indian Affairs, 43, 127
U.S. Bureau of Land Reclamation, 13, 15, 25, 45, 49, 72, 142–144, 169
U.S. Bureau of Public Roads, 28, 32–33
U.S. Department of Agriculture, 49, 172
U.S. Department of the Air Force, 145
U.S. Department of the Army, 145
U.S. Department of Housing and Urban Development, 24
U.S. Department of Transportation, 5
University of Texas, 29, 31–32
Utah Department of Transportation, 32, 128
Uzan, Jacob, 39, 75

Van London, W. J., 62, 93
Van der Muerwe, C. F., 2, 74
Vedros, Philip, 33
Vegetation, 60, 66, 76 (*See also* Tree roots)
Venezuela, expansive soils in, 3, 20, 24, 75
Vermeer trencher, 109, 113, 124, 151, 159
Vertical moisture barriers, 177–181
Vertical placement, 5–6, 62–64, 149, 151–154
Very light density geomembranes, 144
Victoria, Texas, 90
Victoria Roads, 25, 38, 63, 78–79, 83, 130–137, 139, 153–154, 160, 168–169

Waco, Texas, 44, 96–97
Walsh, 58

Walsh procedure, 58, 91
Wardle, 58
Waterways Experiment Station, 2, 24, 33, 43, 49, 94, 172
Watt, Gordon, 31
Webs, 10
West Industries, 13
Wetting boxes, 79
Whitman, R. W., 28, 49
Williams, A. A. B., 39, 74
Wiseman, Gdalyah, 18, 20–21, 25, 34–35, 39, 47, 49, 59, 75
Woodward, J. R., 49
Wooltorton, D., 4, 20–21, 26, 47, 49, 56–57, 70, 91
Woven fabrics, 11

Wray, Kent, 56, 58, 60, 63, 73–76, 89, 162
Wyoming Department of Transportation, 6, 15, 32, 53, 64, 121–126, 139, 152–153, 157, 160, 167–169, 183, 189

Xeriscaping, 66

Yarns, 11, 13
Yazoo clay, 61
Yemec trencher, 159

Zeitlin, Joseph, 2, 39, 75
Zimbabwe, expansive soils in, 2–3

## About the Author

Malcolm Steinberg, P.E., F.ASCE, is an internationally recognized lecturer and expert on expansive soils. The director of an El Paso consulting firm that specializes in soils and foundation engineering, he is currently a District 15 director for the American Society of Civil Engineers and is a member of several transportation research board geomembrane committees. For nearly 40 years, he was an engineer with the Texas Highway Department and the author of many technical papers. This is his first book.